The New York Times
BOOK OF
GENETICS

Other books in the series:

The New York Times Book of Archeology
The New York Times Book of Birds
The New York Times Book of the Brain
The New York Times Book of Fish
The New York Times Book of Fossils and Evolution
The New York Times Book of Mammals
The New York Times Book of Natural Disasters

The New York Times
BOOK OF
GENETICS

EDITED BY
NICHOLAS WADE

THE LYONS PRESS
GUILFORD, CONNECTICUT
An imprint of The Globe Pequot Press

10 9 8 7 6 5 4 3 2 1

Designed by Joel Friedlander, Marin Bookworks

The Library of Congress Cataloguing-in-Publication Data is available on file.

Contents

8   The Latest from the Field

Appendix: A Guide to the Language of Biology

Introduction

GENES ARE THE SOFTWARE of the living cell, the instructions that specify everything it does. Genetics, the study of genes, is going through a period of remarkable progress, yielding knowledge that reaches to the roots of human existence and that may in time change many lives.

The notion of inheritable units was first inferred by the abbé Mendel in the nineteenth century but the modern era of genetics, the study of genes at the level of their molecules and chemistry, began only in 1953 with the discovery of the structure of DNA.

It has taken biologists many years since then to develop appropriate tools for manipulating and understanding DNA. One decisive step was the ability to cut DNA in specific places and move genes from one organism to another. Another technique, known as PCR, enables biologists to amplify any known segment of DNA, producing many identical copies for further study. In the early 1990s, the first DNA sequencing machines became available, allowing the sequence of units in DNA, the chemical letters that hold its genetic information, to be worked out in lengths of 500 or so units at a time.

These and other powerful techniques have contributed to the growing momentum of genetic discovery. But despite the amount that has been learned, genetics could probably be said to be still in its gestational stage. Biologists are still gathering information; making use of it requires development of another set of techniques, which for the most part are not yet in place. Gene therapy, for example, the idea of replacing a defective gene in an individual with a normal version, has been talked of for years but is not yet feasible.

Still, the reach of the new knowledge is evident. Once biologists have the full parts list and operating instructions of the living cell, they will gain deep insights into many diseases. The genetics of the cancer cell are already understood far more clearly than before, and effective treatments will surely one day emerge from this fundamental understanding.

Many inherited diseases, such as Huntington's disease and certain hereditary forms of breast cancer, have now been traced to defects in single genes. It is probably only a matter of time before some of these defects can be corrected.

DNA yields historical as well as medical information. There is only one tree of evolution, and every gene bears in its sequence of chemical letters the traces of its descent from an earlier ancestral gene. The story of human evolution is written in human DNA with a remarkable degree of detail, even including the size of ancestral human populations and the dates of their migrations round the globe.

Driving and unifying the exploration of the genetic universe is the Human Genome Project, an ambitious program now halfway through its 15-year course. The goal of the project is to decode the order of the three billion letters of human DNA by the year 2005. Along with human DNA, the project also aims to sequence the DNA of the smaller animals whose genes have already been much studied in the laboratory, such as the roundworm, the fruit fly and the mouse. The DNA or genomes of these animals contain many genes related to those of humans, and the role of such genes is often easier to work out in experimental animals. The smaller genomes thus help to interpret the human genome.

By changing genes, a biologist can in principle improve existing species and design new ones, achieving in an instant a feat that takes evolution thousands of years. In the case of humans, there is no real controversy about somatic gene therapy, the correction of defective genes in the ordinary tissues of the body. Such a change is like surgery, except on a molecular scale, and does not outlive the patient.

Far more controversial is the impending ability to make genetic changes to germline cells, meaning the egg or sperm. Such changes would propagate to the individual's descendants and last indefinitely. It can be argued that germline corrections of clearly defective genes, such as those that predispose toward cancer or mental illness, would be an undisputed blessing. However, another category of conceivable genetic changes involves those that might be called cosmetic—genes that enhance desirable qualities such as health, strength, beauty or intelligence.

Opponents therefore argue that once one category of germline changes is accepted, it might be hard to draw the line at the next. Before long, all kinds of genetic enhancements would be on offer, bequeathing some novel problems to future generations.

Is germline engineering ethical? If ethical, is the technique open to abuse? It is reasonable to distinguish germline engineering, a voluntary choice of genes that parents might wish to select for their children, from eugenics, the sequestration or murder of people with genes deemed undesirable by governments?

Even if germline engineering were to proceed in a politically acceptable context, it would lead in principle to humans taking over control of human evolution, a step likely to meet objections whether on ethical, religious or biological grounds. Deliberate control might be an improvement over the random forces of evolution that have shaped the human design hitherto. On the other hand, evolution has not done so bad a job so far, and there is a certain presumption in assuming the responsibility of doing better.

These issues are at present mere speculation but will not necessarily remain so forever. The articles selected here, originally written for the science section of the *New York Times*, give a snapshot of where genetics now stands. Collected in book form, they may afford readers a glimpse of the vast reach of the technology that is now unfolding.

The articles are intended to be self-explanatory, but for readers who would like a more systematic explanation of how genes and cells work, a brief outline is offered in an appendix.

I thank Lilly Golden of the Lyons Press for proposing the idea of this book.

—NICHOLAS WADE, Fall 1998

1

GENETICS AS HISTORY

One of the most fascinating aspects of genetics is that an organism's DNA is more than a program for telling its cells how to operate: It is also an archive of the individual's evolutionary history.

From the archive, all kinds of information can be extracted. By comparing DNA among people of different ethnic groups, biologists can infer much about humans' origins in Africa and their subsequent migrations throughout the Old and New Worlds. They can date the migrations and even estimate the size of the populations involved.

This cornucopia of information arises from the record of evolutionary change that is embedded in DNA. Genes change slowly over time, rather like words change their spelling and pronunciation. The changes in genes are called mutations, a mutation being any change in the sequence of chemicals letters in DNA. Thus the genes that make insulin, say, have a similar but not identical sequence of DNA letters in whales, humans and apes, reflecting each gene's evolutionary descent from an ancestral insulin gene.

To pursue the analogy with linguistics a little farther, the ancient Greek word *episkopos* has mutated over the years into many different forms, with its *ps* changing to *v* or *b* according to known linguistic rules. The mutations differ from one language to another, just as the same gene is often spelled differently in different species. In Italian the word has become *vescovo,* in French *évêque,* in Spanish *obispo,* and in Middle English *bisceop.* In all its changing forms, *episkopos* has retained its meaning of "bishop."

Just as a linguist from this example could draw up a family tree of these various forms, so biologists can infer family trees of related genes. Genes from different species that are related in this way are said to be homologous, an important concept in studying evolutionary origins.

The family tree of a homologous set of genes shows where each branch forks off from the main trunk but not when it did so. The date of the branching points can be estimated by counting the number of changes that have occurred in the DNA. The changes have a variety of natural causes, such as radiation, chemical decay and copying errors made as DNA is duplicated in cell division. The number of changes reflects the amount of time that has elapsed.

Molecular clocks, as time-measuring pieces of DNA are known, are tricky to read and biologists often disagree as to what time they are telling. One problem is that different regions of DNA tick at different rates. DNA in the genes ticks very slowly because if a gene is altered its protein product may not work properly, with fatal effects for the organism.

The best kind of molecular clock is found in the DNA which is not in the genes. In animals, most DNA does not code for genes and a lot of it has no apparent function. In humans, noncoding DNA accounts for 97 percent of the genome. The genes are strung along the DNA like small islands in a vast ocean.

The noncoding DNA that lies between the genes gives good dates because it is free to change or mutate at a steady rate without any risk that the mutations will cause a change in a protein.

DNA gets thoroughly shuffled all along its length when egg and sperm cells are formed, which is why one individual differs from another. The shuffling process can confound molecular clocks. Luckily, there are two kinds of DNA which are exempt from shuffling.

One is the DNA on the Y chromosome. During the shuffling process, called meiosis by biologists, the Y chromosome exchanges DNA at its very tips with its sister chromosome, the X, but the main body of Y does not mingle with the X (if it did, the male-determining gene would escape into the female genome, wreaking havoc with the natural sex ratio).

The noncoding region of the Y chromosome thus makes a wonderful natural archive for tracking the ancestry of males back to the original Adams in the founding population of the human species.

A quite different kind of DNA leads back to one of the first Eves. Mitochondrial DNA, as it is known, comes from the tiny, bacteria-like organelles that supply the human cell's energy needs. The mitochondria were once free-living bacteria that became enslaved in the dawn of time by animal and plant cells. They have lost most of their genetic functions, save those that handle the energy metabolism needed by their captors.

The DNA in the nucleus of a fertilized egg is supplied equally by both parents but mitochondria and their DNA come only from the egg. A father's name, in some societies, may go to all his progeny, but the mother's mitochondria, a much more lasting gift, are bequeathed to all hers. The noncoding regions of mitochondrial DNA serve as a useful molecular clock for dating human prehistory, particularly since mitochondrial DNA tends to mutate rather faster than nuclear DNA and thus to provide a finer scale of measurement.

To People the World,
Start with Five Hundred

AS FEW AS 500 OR SO PEOPLE, trekking out of Africa 140,000 years ago, may have populated the rest of the globe. These estimates are derived from a novel kind of archaeology, one that depends not on pick and shovel but on delving into the capacious archive of the human genome.

Dates and numbers based solely on genetic evidence are unlikely to be fully accepted until historians and archaeologists have had their say. But they afford a glimpse of the rich historical information embedded in the DNA of each human cell. Because of rapid methods for sequencing, or reading off, the chemical letters of DNA, geneticists are gathering reams of data about human populations. But methods of interpreting the information are still a work in progress.

At a conference on human evolution in October 1997 at the Cold Spring Harbor Laboratory on Long Island, population geneticists reported new analyses confirming that the origin of the human species is to be sought in Africa. They also discussed other inferences that can be teased out of genetic data, like estimates of the size and location of ancestral populations and the timing of the migrations out of Africa.

The geneticists' calculations depend on a number of assumptions, and tend to yield dates that have wide margins of error, a source of frustration to others at the conference. "Why should we worry about their dates?" one archaeologist said to another. "It's they who should worry about our dates."

Still, a method that from a few drops of blood can reach into the dawn of human history is hard to ignore, whatever its present imperfections.

Overall, the genetic variation found among people is very small compared with that among most other species; humans evolved too recently to have accumulated any significant amount of genetic change or mutations.

Still, people are far from being clones and there is a lot of variation for geneticists to track, much of which was present at the emergence of the species.

Movements of populations have been studied in the past by comparing the cell's working parts, the proteins whose structure is coded in DNA. But a coding region of DNA cannot change very much, since mutations will alter the protein it specifies, often with fatal consequences for the individual. Most of the genome, however, is composed of noncoding DNA, where mutations make no difference to the individual since most noncoding DNA has no evident purpose. Mutations in noncoding DNA are ideal for the population geneticist, since they accumulate at a fairly regular rate, yielding the best data as to the diversity and age of different populations.

One result that stands out from genetic samples of people around the world is that sub-Saharan African populations possess greater genetic diversity than non-Africans. Non-Africans retain just a subset of this diversity, as would be expected when a smaller group breaks away from a founding population, taking only a sample of the full range of genetic variation.

Dr. Mark Stoneking, a population geneticist at Pennsylvania State University, estimated at the conference that the non-Africans had split away from the main human population in Africa 137,000 years ago, give or take 15,000 years. His findings were based on sampling 34 populations around the world and analyzing genetic elements called Alu insertions, small and apparently useless pieces of DNA that have gradually spread throughout the human genome.

The substantial diversity of Alu insertions among African populations suggested that Africans maintained a larger population size through the prehistorical period than those who emigrated, Dr. Stoneking said, since small populations tend to lose their genetic diversity over time.

The original human population is thought to have numbered a few thousand individuals. Studies of mitochondrial DNA, a special category of genetic material that is inherited just through the mother's line, have put the founding human population at a mere 10,000 individuals. Using genetic markers from the chromosomes, Dr. Sarah A. Tishkoff, an evolutionary biologist at Penn State, said she and her colleagues had calculated that the long-term early population was considerably larger, from 23,000 to 447,000 individuals.

The home of the ancestral human population in Africa is not yet known but some signs point toward East Africa. The Turkana people of Kenya show the greatest diversity of any known group in their mitochondrial DNA, said Dr. Elizabeth Watson of Massey University in New Zealand. This type of genetic material is exempt from the shuffling that creates new individuals and changes only by collecting mutations over time.

Dr. Watson said that other peoples of East Africa also had high diversity in their mitochondrial DNA, and that the region of the highest diversity was usually indicative of a species' place of origin. Since fossil remains of early human ancestors have been found at Lake Turkana in Kenya, her proposal is plausible, other experts said, although more diverse African groups may turn up after wider sampling.

Dr. Tishkoff believes that the African and non-African populations may have separated in several stages. She has found that the Falashas, Ethiopian Jews who now live in Israel but are believed to be like other Ethiopians genetically, show genetic diversity intermediate between that of Africans and non-Africans, suggesting they may be descendants of a group that moved away from the main sub-Saharan population and lived perhaps in northeast Africa. Today's non-African populations could have originated from the northeast African group, carrying off even lesser genetic diversity.

There may have been more than one migration out of Africa, but if so all probably came from northeast African since the pattern of variation among non-Africans suggests that they all descend from a single ancestral gene pool, Dr. Tishkoff said.

The genetic diversity of non-Africans is so much smaller that the group leaving Africa could have consisted of as few as 200 to 500 individuals, Dr. Tishkoff and colleagues estimated, although, of course, the actual numbers may have been greater.

There seems to have been one principal migration of modern humans out of Africa, but geneticists have picked up traces of at least one earlier exodus. Dr. Stoneking believes from his Alu insertion study that peoples now found only in parts of Australasia are closely related to the ancestral human population and may have reached the far Pacific through a tropical route, to be largely displaced by a later wave of people coming through Central Asia.

A family tree of human populations can be drawn up from analysis of mitochondrial DNA, which because it accumulates mutations quite rapidly is useful for studying recent evolutionary events. The pattern of mutations in mitochondrial DNA can be traced back to an ancestral mutation pattern, whose date can be estimated by various means. Dr. Douglas C. Wallace of Emory University in Atlanta said that the age of all African mitochondrial DNAs could be set at 130,000 to 170,000 years ago, that of Asians at 50,000 to 70,000 years and the root of the European lineages at 40,000 to 50,000 years ago.

Native Americans, by the evidence of their mitochondrial DNA, fall into four genetic groups, three of which can also be found in most present-day Siberians, Dr. Wallace said.

Population geneticists have started to analyze the male, or Y, chromosome, which like mitochondrial DNA is a useful genetic marker because it largely escapes the reshuffling of genes in each generation. The reason is that the Y chromosome exchanges genetic information just at its tips with its counterpart, the X chromosome, leaving the main body of the Y to change only by successive random alterations in its DNA. As with mitochondrial DNA, plotting these random mutations yields a tree with the oldest mutation at the trunk and younger mutations forming successive branches.

Mitochondrial DNA and the Y chromosome should tell more or less the same story of population history unless men and women did significant travel without each other. From studying the Y chromosome of 1,500 men around the world, Dr. Michael Hammer of the University of Arizona said the family tree of present mutations started around 185,000 years ago, a date similar to that of the ancestral mitochondrial DNA.

But unlike the mitochondrial DNA tree, which mirrors the migration of people out of Africa and around the world, Y chromosome DNA shows evidence of a male migration from Asia back into Africa. This finding implies "that males dispersed more than females during these major population movements," Dr. Hammer said.

The new findings described at the conference, buttressed by the recent sequencing of mitochondrial DNA from the fossil bone of a Neanderthal, support the belief of many paleoanthropologists in the out-of-Africa theory. This is the view that modern humans emerged from Africa and displaced, without interbreeding, the archaic hominids like Neanderthals that had populated the world during earlier emigrations. Dr. Milford Wolpoff of the

University of Michigan staunchly upheld his alternative view that modern humans evolved from Neanderthals in Europe and from other archaic hominids in different parts of the world. At least for the moment, however, he seems to be in an embattled minority.

Remains of anatomically modern humans have been found from as long as 200,000 years ago in Africa but some archaeologists, like Dr. Richard G. Klein of Stanford University, distinguish between anatomically modern and behaviorally modern people. A suite of new artifacts, like the making of art objects, appears in the archaeological record in Africa 50,000 years ago and a few thousand years later in Asia and Europe. "There was some kind of Rubicon that was crossed 40,000 to 50,000 years ago," Dr. Klein said, suggesting that the Rubicon was a neurological change that permitted the development of language.

Another archaeologist, Dr. Stanley Ambrose of the University of Illinois at Urbana, agreed with Dr. Klein on the date of the change but said that language had developed much earlier. He argued that the behavioral changes seen at 50,000 years ago reflected a population recovery after an adverse climate change, particularly a deep global cold snap caused by the gigantic volcanic eruption of Mount Toba in Sumatra 73,500 years ago.

Population genetics is at its most fascinating in exploring the origin of humans, but can also be brought to bear on more recent events like the peopling of the Americas or the prehistory of populations in Asia. During an expedition across Central Asia last summer, Dr. Spencer Wells of Stanford University collected blood from speakers of 32 different languages in the region. Analyzing the DNA in his laboratory, he found that many Turkish speakers carry a harmless mutation that first arose among the Indo-Iranian peoples who lived north of the Caspian Sea some 5,000 years ago, as well as another genetic marker found among Mongoloids.

A southward migration of the Indo-Iranian speakers into India began the spread of the Indo-European languages, which include English. There must also have been an eastward migration, Dr. Wells believes, that reached the Altai Mountains north of Mongolia, and from a fusion with Mongoloid peoples of the Altai, the Turkish-speaking peoples emerged.

—NICHOLAS WADE, November 1997

Gene of Mideast Ancestor
May Link Four Disparate Peoples

SEVERAL THOUSAND YEARS AGO, somewhere in the Middle East, there lived a person who bequeathed a particular gene to many present-day descendants. But these millions of now distant relatives could not convincingly be called one big happy family. They include Jews, Arabs, Turks and Armenians.

The gene, a variant of a gene that controls fever, has come to light because it causes an unusual disease called familial Mediterranean fever in people who inherit a copy from both parents. The gene's presence among a surprising group of populations hints at the rich archaeology that lies buried in the human genome, once geneticists and historians have learned how to interpret it.

Two rival teams of scientists in France and the United States have been racing to isolate the gene for a year. The race has finished, with the American team announcing its finding in the journal *Cell,* the French team in *Nature Genetics.*

The American team has named the gene pyrin, from the Greek word for fire, after its role in fever; the French team calls it marenostrin after the Latin "Our Sea," a Roman phrase for the Mediterranean. The race could be considered a dead heat, although the American team has recovered the whole gene, the French team just a major portion.

People who inherit a single variant copy of the fever gene from one parent and a normal copy from the other parent have no sign of the disease. They are so numerous, constituting up to 20 percent of certain Jewish and Armenian populations, that carrying one copy is assumed to confer some significant benefit, like a greater resistance to disease.

In people with two copies, however, the immune system goes into overdrive at inappropriate moments, causing bouts of severe fever. The scientists who have analyzed the fever gene and its variants say they now understand why.

The normal gene specifies a protein that from its design motifs looks as if it is meant to slip into the nucleus of the cell and switch genes on or off. Since the gene is active only in a special class of white blood cells, its usual duty seems to be to control the cells' activity and rein them in when the threat of infection has passed. The white blood cells defend against infections and often cause fever in doing so.

The new findings, in portraying the exact genetic anatomy of the normal gene and its variant forms, give a strong clue as to why the variant versions have the effects they do. The variant forms have mutations, or changes of a single DNA letter, in the region of the gene assigned to the switching function. Presumably the mutations make the gene's protein inefficient in its duty of restraining the white blood cells.

The historical significance of the finding lies in the genetic relationship it implies between populations that have been separate for many hundreds of years. For example, the variant form of the gene found in North African Jews, Iraqi Jews and Armenians is the same, carrying both the same mutation and a pattern of 11 other genetic changes, all harmless. Although single genetic changes can arise independently, the presence of so many together in the same combination points strongly to a "founder," or single ancestor, as the original source of the variant gene.

A second variant form of the gene, according to the American team, is shared by Iraqi Jews, Ashkenazi Jews, the Moslem Druze sect and Armenians. The two variants are similar and probably derive from the same founder.

The Americans write that the mutations are "very old" and that they suggest "common origins for several Middle Eastern populations."

Dr. Daniel L. Kastner, a member of the team, said the original possessor of the variant gene probably lived several thousand years ago and certainly less than 40,000 years ago, according to a formula that relates the average length of a shared genetic segment to the number of generations that have passed. Dr. Kastner said the founder's gene may have spread through a population in the Middle East that existed before Jews, Armenians and Arabs became distinct peoples.

He also noted that the variant fever gene established a common genetic lineage between Ashkenazi Jews and Iraqi Jews, even though the two communities have been separated since the Babylonian Captivity that began in 597 B.C. Many Jews from the ancient community in Iraq now live in Israel.

The French team has detected the main variant in Jews and Arabs from North Africa and in Turks and Armenians. Dr. Jean Weissenbach of the gene laboratory Genethon, a member of the French team, said that the variant gene was ancient but that an exact date of its origin could not be calculated.

Experts in Middle Eastern history and linguistics said they knew of no historical event to link the four populations in which the variant fever gene has been found, although three—Arabs, Jews and Armenians—are related geographically, having originated in the Middle East. The ancestral Armenian homeland is around Lake Van in Turkey. The Seljuk Turks invaded from Central Asia in the 11th century, and they absorbed many of the local inhabitants.

Familial Mediterranean fever is rare in the United States. Patients often endure years of misdiagnoses. Once the disease is recognized, an effective drug, colchicine, is available. Now that the DNA sequence of the variant gene is known, an accurate test can be made.

"This will really help to diagnose the disease in the very early stages and to start the colchicine treatment as soon as possible," Dr. Weissenbach said.

The American-led team includes scientists from Israel and Australia. The French team is from Genethon in Evry and two other laboratories.

—NICHOLAS WADE, August 1997

Finding Genetic Traces
of Jewish Priesthood

IN AN UNUSUAL MARRIAGE of science and religion, researchers have found biological evidence in support of an ancient belief: Certain Jewish men, thought to be descendants of the first high priest, Aaron, the older brother of Moses, share distinctive genetic traits, suggesting that they may indeed be members of a single lineage that has endured for thousands of years.

The men are known as Jewish priests, a designation that since the time of Aaron 3,300 years ago has been passed down through the generations, exclusively from fathers to sons. The only way to become a priest is to be born the son of one. They differ from rabbis, though a priest may choose to become a rabbi. Historically, certain blessings and rituals could be performed only by priests, and some congregations today still follow that tradition.

Many priests have the surname Cohen or Kohen, which in Hebrew means priest, or related names like Kahn or Kahane. Those with other surnames generally have the words *ha'kohen,* for "the priest," inscribed on their gravestones, sometimes with an image of hands raised in a characteristic gesture of blessing. Even in families where priests no longer perform the traditional religious duties, knowledge of the heritage is often preserved.

It was the patrilineal nature of Jewish priesthood that piqued the curiosity of a research team from Israel, England, Canada and the United States. Knowing that another bit of a man's identity is also passed strictly from father to son—namely, the Y chromosome, which carries the gene for maleness—they set out to determine whether that chromosome might carry special features that would link the priests to each other and set them apart from other men, confirming the priests' unique paternal lineage.

"I think anybody who knows the biblical story about Aaron and this tradition of the priesthood going from father to son, and is aware that the Y chromosome is inherited in the same way, would think of this question," said Dr. Michael Hammer, a geneticist at the University of Arizona in Tucson, and an author of a report about the priests in the journal *Nature*.

A unique aspect of the Y chromosome that lends itself to this sort of research is that the Y does not swap stretches of DNA with other chromosomes. Changes that occur in the Y tend to persist in a lineage over time, and, Dr. Hammer said, "We can use that to interpret historical events." In a study published in 1995, he and his colleagues used segments of the chromosome to suggest that all men living today can be traced back to a common ancestor 188,000 years ago.

The subjects of the current study were 188 Jewish men from Israel, North America and England. The researchers did not rely on surnames to identify priests, but instead asked the men if they had been told they were priests. Sixty-eight had, and the rest identified themselves as "Israelites," a term used to describe laymen. (Men who said they were Levites, members of a different priesthood, were omitted from the study.)

The researchers obtained Y chromosomes by extracting them from cell samples scraped from the men's mouths. They studied two sites, or markers—known variable regions of DNA—to find out whether the priests and Israelites differed.

They did. Only 1.5 percent of the priests, as opposed to 18.4 percent of the laymen, had the first marker. The other marker, which could appear in five different forms, tended to occur most often in one version in the priests. Fifty-four percent of the priests had this version and 33 percent of the others had it.

"The simplest, most straightforward explanation is that these men have the Y chromosome of Aaron," said Dr. Karl Skorecki, a coauthor of the report who conducts genetic research at the Technion-Israel Institute of Technology, in Haifa. "The study suggests that a three-thousand-year-old oral tradition was correct, or had a biological counterpart." There are at least 350,000 priests around the world today with that same chromosome, he and his colleagues estimate, about 5 percent of the Jewish male population. They are all related, Dr. Hammer said, and could be considered distant cousins on their fathers' side.

"It's a beautiful example of how father-to-son transmission of two things, one genetic and one cultural, gives you the same picture," Dr. Hammer said.

The study also supports the idea that the priesthood was established before the world Jewish population split into two major groups a thousand years ago, as a result of migrations. The marker findings in the priests were similar in Ashkenazic and Sephardic Jews, indicating that the priesthood antedated the division.

Asked to comment on the study, Dr. James Lupski, a medical geneticist at the Baylor College of Medicine in Houston, said: "It's amazing to think how you can use these technologies to investigate history and evolution. They took a very interesting approach that will certainly be useful for studying the Y chromosome. And a report like this is going to stimulate interest, stimulate other groups around the world to confirm it in a different way."

Dr. Hammer said he did not know whether the chromosome testing used in the study would be of interest to anyone other than scientists. But, he said, "I do know someone named Cohen who said he'd be interested in having the test, just to find out if he was really a priest." At this point, the test could suggest that a man was a priest, but not prove it. It could, however, rule out the possibility with a high degree of certainty.

"It could say your DNA is not consistent with patrilineal descent from a common ancestor, Aaron," Dr. Skorecki said. "Whether the religious community would accept that as grounds for exclusion is not an issue I'd want to get into. It's for the rabbis to debate."

Rabbi Aaron Panken, of the Hebrew Union College Jewish Institute of Religion in Manhattan, said: "There's a lot of danger in this for religious fanatics to go off in different directions. It could become a tool for fundamentalists to try to weed out who is not a cohen, and that would be troublesome."

In addition, Rabbi Panken said, because priests were traditionally banned from marrying divorced women, he could imagine fundamentalist groups demanding DNA testing before permitting any man to marry a divorced woman, to make sure the man was not a priest.

"It would also concern me if we began to look backwards," he said, "romanticizing the hereditary model of priestly leadership. Do we want a hereditary leadership pattern in the Jewish community? We haven't had that in two thousand years."

—DENISE GRADY, January 1997

Research Team Takes Big Stride in the Mapping of Human Genes

REACHING A SIGNIFICANT GOAL in the project to explore the mysteries of human genes, biologists have completed a high-resolution map of the X chromosome, one of the pair that determines whether a baby is a boy or a girl.

The map consists of a set of identifiable milestones at frequent intervals along the chromosome, which is a giant molecule of DNA some 160 million chemical units in length.

Although rough maps of the X and other chromosomes have been made before, this is the first time that any chromosome has been mapped to the level of detail set by the Human Genome Project, a $3 billion effort to describe the human genetic blueprint completely, said Dr. Eric Green of the National Human Genome Research Institute.

The map, which took 10 years to complete, is the work of 25 biologists at the Washington University School of Medicine in St. Louis. The team's work, directed by Dr. David Schlessinger, was reported in the journal *Genome Research*.

CLOSE-UP

The X Chromosome, in Detail

Researchers have completed a map that shows the X chromosome — one of the pair that determines a person's sex — in far more detail than ever. If it were a road map of a 2,100-mile highway, it would have an identifying marker every mile.

New X chromosome map

MILLIONS OF BASE PAIRS
- 0
- 15
- 30
- 45
- 60
- 75
- 90
- 105
- 120
- 135
- 150
- 160

■ NEW FINDINGS
The researchers located a large region on the chromosome where DNA remains intact as it passes from one generation to another.

■ POTENTIAL FINDINGS
The new map will allow scientists to hunt for disease genes on the X chromosome, which is associated with many inherited disorders.

BASE PAIR

HIGHER RESOLUTION The X chromosome has 160 million chemical units, or base pairs; the new map has a signpost for every 75,000 pairs

Sources: Dr. David Schlessinger/Washington University School of Medicine; Bruce Alberts et al., editors, "Molecular Biology of the Cell"

N.Y. Times News Service

21

The X chromosome may to many people connote femaleness, since women have a pair of X chromosomes and men have an X and a Y.

But it is of intense interest to geneticists beause of a reason that has to do with males: the fact that many conditions caused by defective genes on the X chromosome, like hemophilia and color blindness, turn up only in men. The reason is that women can often compensate for a defective X chromosome gene if the counterpart gene on their second X chromosome is in working order.

Dr. David Nelson, an X-chromosome expert at the Baylor College of Medicine, described the map produced by Dr. Schlessinger's team as a "tour de force."

Dr. Huntington F. Willard, a geneticist at Case Western Reserve University, said the map "will greatly speed up searches for X linked–disease genes and efforts to understand chromosome biology."

The map will enable other laboratories to sequence, or chemically identify, the DNA units in the regions between the map's milestones, which are themselves short sequences of DNA about 75,000 units apart. By filling in all the gaps between the milestones, the full DNA sequence of the X chromosome can be determined, giving a complete description of the human genetic blueprint in at least one of the 24 chromosomes that make up the human genome.

The map is also a gift to researchers studying X-linked diseases. Without waiting for the full DNA sequence, they can now locate a gene's likely position on the X chromosome more rapidly by identifying the milestones that are most closely inherited along with the gene of interest.

Even though Dr. Schlessinger and his team did not sequence any genes on the X chromosome, they were able to infer from the DNA in their milestones that the genes are not spread out evenly but instead are bunched together mostly in five gene-rich regions. They also discovered a large segment of the chromosome that is very loath to recombine with its partner in the gene-shuffling dance and ensures that individuals will differ from their parents.

Other geneticists said these findings confirmed two emerging features of the structure of human chromosomes: that the genes are arranged in a clumpy way, with lengthy, apparently desert regions of DNA in between, and that some regions of the chromosomes are more prone to recombine than others.

The reasons for these major features remain to be explained but presumably hold a key to the principles on which chromosomes are organized.

The finding of gene-rich pockets on the X chromosome raises an issue of strategy for the four or five groups interested in sequencing it. Some biologists want to sequence the entire chromosome as soon as possible, while others argue that it would be better to sequence the gene-rich regions first and save the gene-poor regions for later, when the cost of sequencing will be lower.

The availability of the map should also help unravel some other puzzles of the X chromosome's biology. Militant feminists and male chauvinists alike may be surprised at geneticists' belief that the X and Y chromosomes were once identical—many millions of years ago. But when the Y acquired the gene that determines male sex, the X and the Y started to follow different evolutionary paths.

The Y, for the most part, stopped mingling its genes with those of the X, and, for lack of the rejuvenation conferred by recombination, its genes grew obsolescent. Because they could be shed with impunity, the Y grew shorter and shorter.

Its shrinkage seems likely to continue, to the vanishing point. But geneticists are not yet declaring men candidates for the endangered-species list.

They believe a different mechanism of sex determination will eventually evolve, maybe with a single X causing maleness, or with the male-determining gene jumping to a new pair of chromosomes and the story of X and Y starting all over.

While the Y chromosome is unstable and flighty, at least on the evolutionary time scale of millions of years, the X chromosome, in contrast, is the most conservative of all the chromosomes. Genes tend to jump around between the other chromosomes, but those on the X stay put to such an extent that mice and humans have much the same set of genes on their X chromosomes even though the two species have followed separate evolutionary paths for 80 million years.

The X chromosome is also special because one of the pair in every woman's cells is somehow inactivated. It is because the inactivation is random, with either the maternal or paternal X chromosome being switched off in each cell, that women tend to be protected from X-linked diseases.

The mechanism of inactivation is not well understood but may have something to do with the chromosome's evolutionary stability. It also seems that the older parts of the chromosome accept inactivation better than do the younger regions.

"This map will help provide data to settle these questions," said Dr. Willard, the geneticist at Case Western Reserve.

Dr. Green, of the National Human Genome Research Institute, attributed part of the success of Dr. Schlessinger's team to techniques of gene mapping developed at Washington University several years ago.

Using the same techniques, Dr. Green said, he has mapped human chromosome 7 to equal resolution. But he noted that Dr. Schlessinger's result was the first to be published.

—NICHOLAS WADE, March 1997

Male Chromosome Is Not a Genetic Wasteland, After All

WHY THE Y CHROMOSOME? The question can now be answered, after the discovery of almost all of its genes and the emergence of a pattern that explains how the Y chromosome evolved.

The new findings may also help toward understanding the sources of male infertility, the cause of half the childlessness that affects some 10 percent of American couples. Defects on the Y chromosome affect only men, who have an X and Y chromosome in each cell of their bodies, while women have two Xs.

The Logic of the Y Chromosome

Diagram of the Y chromosome shows its two classes of genes. The X-homologous genes (black), so called because they have close counterparts on the X chromosome, are general housekeeping genes, switched on in many cells. The other class (gray), which includes the male-determining gene SRY, is switched on only in the testis. Third line shows result when given genes are deleted.

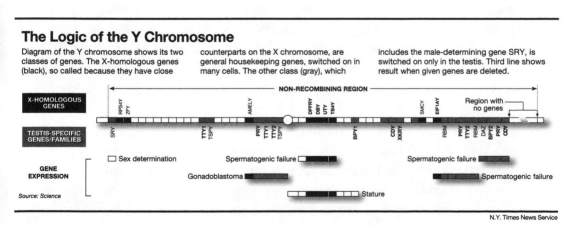

Source: Science

N.Y. Times News Service

Chromosomes, visible under the microscope as puffy X-shaped blobs, are nature's answer to the hard disk drive: they are high-density genetic storage devices organized along random access principles. The 70,000 human genes are packed into 23 pairs of chromosomes but are distributed, as far as anyone can tell, with no rhyme or reason as to which chromosome a gene is assigned to.

The one exception is the Y chromosome, which has very few genes on its main body, apart from the male-determining gene that switches the fetus away from the default human condition of femaleness.

It is strange that this bastion of maleness should be a wasteland, genetically speaking. Until recently, only eight genes had been found on it, compared with the several thousand packed into each of the other 45 chromosomes, including the Y's counterpart, the X chromosome.

Believing there must be something more in this desolate genetic terrain, two biologists at the Whitehead Institute in Cambridge, Massachusetts, Dr. Bruce T. Lahn and Dr. David C. Page, made a thorough search of the Y chromosome and have now turned up 12 more genes. That makes 20 known genes on the main body of the Y, which they believe is close to the full inventory.

Knowing roughly what these genes do, the Whitehead scientists have been able to reconstruct the likely history of how the Y chromosome got to be the way it is. Their findings and theory were reported in the journal *Science*.

In the beginning, when devising two sexes, nature needed to ensure that the male-determining gene did not sneak into the female's genome or gene set. That is a tougher chaperoning problem than it may sound, because a principal way of creating genetic variety between generations is for pairs of chromosomes to line up and promiscuously trade fragments of genetic material before creating egg and sperm cells. This wholesale shuffling of the genetic deck is called recombination and is a principal reason that children differ from their parents.

To confine the male-determining gene to the male line, the chromosome containing it started not to recombine with its counterpart, except at the very tips, where it has the same few genes as the X chromosome.

But recombination is the genome's tonic for healthy living and all-round fitness. Genes that do not recombine eventually cease to provide useful service to an organism trying to survive in a changing environment. And useless genes tend to get lost. So the Y's decision not to recombine with the X meant that most of its genes became unfit and over the course of millions of years were driven into extinction. The Y has visibly shrunk to just a cut-down version of its sister chromosome.

The Whitehead biologists find that the 20 genes left on the Y fall into two classes. One class has counterpart genes on the X chromosome. Both the X and Y versions of these genes are switched on in many different tissues of the body and play housekeeping roles in the cell's economy. These nine genes are presumably the sole survivors of the Y chromosome's original full complement of genes.

The second class of Y genes is switched on just in the testis and is presumably involved in the manufacture of sperm. Dr. Page views them as male fitness genes, and has reason to believe that over the course of evolution they jumped onto the Y from other chromosomes.

One of these male fitness genes, which has been studied earlier in Dr. Page's laboratory, turns out to be missing in certain men who produce no sperm. The DAZ gene, as it is called (short for "deleted in azoospermia"), seems to be descended from a gene called DAZL (for "DAZ-like") which is found on chromosome 3. The DAZL gene, Dr. Page believes, must have copied itself onto the Y chromosome some 30 million to 40 million years ago, since only in Old World monkeys, and not in animals that split away earlier from the primate lineage, is DAZ found on the Y chromosome.

The other testis-specific genes probably also hitchhiked their way onto the Y chromosome, Dr. Page suggests. Although the chromosome offers an unstable and hostile environment for most genes, it is evidently a sanctuary for male fitness genes. Once they have landed on the Y, these genes produce extra copies of themselves. DAZ, like the other 10 Y genes of its class, exists as a family of multiple copies. The reason for the amplification is unknown.

The two classes, housekeeping and male fitness genes, account for all the known genes on the main body of the Y chromosome and thus explain the one part of the genome where gene content seems to have a rational basis.

"It's a lovely piece of work," said Dr. Huntington F. Willard, an expert on the X chromosome at Case Western Reserve University in Cleveland. "This is the first glimpse we have had that the human genome was thrown together with some forethought."

Dr. Page, who has been studying the Y chromosome for 15 years, described the diminutive chromosome as "a part of the genome I have cared deeply about for some time" and said it was "exciting to find a little pocket of the genome that makes some sense."

Besides male infertility, Dr. Page's findings are also relevant to Turner's syndrome, a condition in women who have only one X chromosome. In normal women, with two Xs in each cell, one X chromosome's genes are permanently switched off so as to give women the same dose of X-based genes as men. Dr. Page has found that there is an exception to this rule: the X-based genes that are the counterparts to the housekeeping genes of the Y. The housekeeping genes remain switched on in both Xs so as to equal the male cell's dose from its X and Y chromosomes.

This suggests, though it has yet to be proved, that the abnormalities seen in Turner's syndrome may arise because the single X chromosome produces only half the needed dose of the various housekeeping genes. The odd collection of symptoms found in Turner's syndrome—short stature, webbing of the neck, infertility and a narrowing of the aorta—may each be caused by insufficient quantities of a different gene. Once these genes are identified, it may be possible to devise drugs that make them more active.

—NICHOLAS WADE, October 1997

Tree of Life Turns Out to Have Surprisingly Complex Roots

FROM YELLOWSTONE PARK to the ocean's abysses, researchers are in hot pursuit of the universal ancestor. Not the sort that is painted in oils and hung proudly in the hallway, but a single-cell creature with few distinctive features save a fondness for living in near-boiling water.

Though the universal ancestor probably lived more than 3.5 billion years ago and was too small to be seen, it was far from contemptible. From this Abraham of microbes sprang the three great kingdoms of evolution, the Prokarya (bacteria), the Archaea and the Eukarya. Plants, fungi and animals, the world's visible life, are slender green shoots at the tip of the eukaryan branch.

Biologists have long aspired to paint a genetic portrait of the ancestor by running the tree of evolution backward, going from its leaves—the living creatures of today—down to the point where all its branches coalesce in a single trunk. Defining the organism that existed at this point, and when and where it lived, might help toward one of biology's major goals, understanding the origin of terrestrial life.

However, the long-standing road map for finding the universal ancestor turns out, in light of new data, to have been misleading, and the map's chief author, Dr. Carl Woese of the University of Illinois, is proposing a new theory about the earliest life-forms.

Working back to the ancestor, an exercise based on the sequence of DNA letters in genes, resembles the way that linguists reconstruct the words of vanished mother tongues from their living descendant languages. Genes that perform the same role in human cells and in bacterial cells, say, may have a recognizably similar spelling of their DNA letters, reflecting the genes'

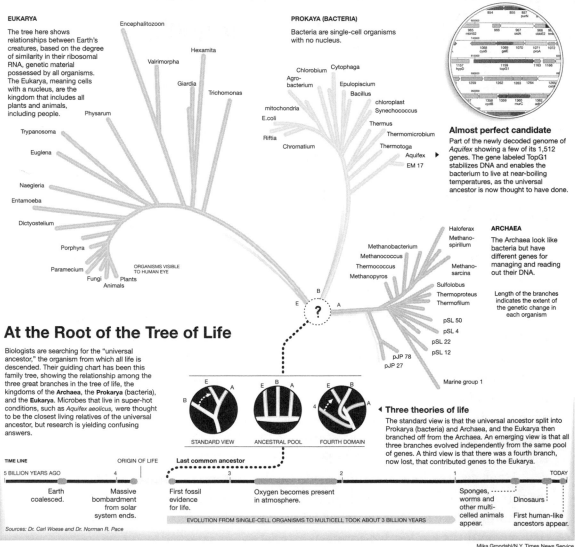

EUKARYA

The tree here shows relationships between Earth's creatures, based on the degree of similarity in their ribosomal RNA, genetic material possessed by all organisms. The Eukarya, meaning cells with a nucleus, are the kingdom that includes all plants and animals, including people.

Encephalitozoon
Hexamita
Valrimorpha
Giardia
Trichomonas
Physarum
Trypanosoma
Euglena
Naegleria
Entamoeba
Dictyostelium
Porphyra
Paramecium
Fungi
Plants
Animals

ORGANISMS VISIBLE TO HUMAN EYE

PROKAYA (BACTERIA)

Bacteria are single-cell organisms with no nucleus.

Chlorobium
Cytophaga
Agro-bacterium
Epulopiscium
Bacillus
chloroplast
Synechococcus
mitochondria
E.coli
Thermus
Thermomicrobium
Riftia
Thermotoga
Chromatium
Aquifex
EM 17

Almost perfect candidate

Part of the newly decoded genome of *Aquifex* showing a few of its 1,512 genes. The gene labeled TopG1 stabilizes DNA and enables the bacterium to live at near-boiling temperatures, as the universal ancestor is now thought to have done.

ARCHAEA

The Archaea look like bacteria but have different genes for managing and reading out their DNA.

Length of the branches indicates the extent of the genetic change in each organism

Haloferax
Methano-spirillum
Methanobacterium
Methanococcus
Thermococcus
Methano-sarcina
Methanopyros
Sulfolobus
Thermoproteus
Thermofilum
pSL 50
pSL 4
pSL 22
pSL 12
pJP 78
pJP 27
Marine group 1

At the Root of the Tree of Life

Biologists are searching for the "universal ancestor," the organism from which all life is descended. Their guiding chart has been this family tree, showing the relationship among the three great branches in the tree of life, the kingdoms of the **Archaea**, the **Prokarya** (bacteria), and the **Eukarya**. Microbes that live in super-hot conditions, such as *Aquifex aeolicus*, were thought to be the closest living relatives of the universal ancestor, but research is yielding confusing answers.

STANDARD VIEW
ANCESTRAL POOL
FOURTH DOMAIN

◄ Three theories of life

The standard view is that the universal ancestor split into Prokarya (bacteria) and Archaea, and the Eukarya then branched off from the Archaea. An emerging view is that all three branches evolved independently from the same pool of genes. A third view is that there was a fourth branch, now lost, that contributed genes to the Eukarya.

TIME LINE

ORIGIN OF LIFE

Last common ancestor

5 BILLION YEARS AGO   4   3   2   1   TODAY

Earth coalesced.
Massive bombardment from solar system ends.
First fossil evidence for life.
Oxygen becomes present in atmosphere.
Sponges, worms and other multi-celled animals appear.
Dinosaurs
First human-like ancestors appear.

EVOLUTION FROM SINGLE-CELL ORGANISMS TO MULTICELL TOOK ABOUT 3 BILLION YEARS

Sources: Dr. Carl Woese and Dr. Norman R. Pace

Mika Grondahl/N.Y. Times News Service

descent from a common ancestor. In one such gene the human-bacterium similarity is as high as 45 percent.

Hope of reconstructing the ancestor from its inferred genes received new impetus in 1995, when the first full DNA, or genome, of a bacterium was decoded. Since then, the genomes of a dozen microbes have been sequenced, including at least one from each of the three main branches of the evolutionary tree.

The three kinds of genome offered a broad basis for triangulating back to the ancestral genome. But the emerging picture is far more complicated than had been expected, and the ancestor's features remain ill defined, though not wholly elusive. "Five years ago, we were very confident and arrogant in our ignorance," said Dr. Eugene V. Koonin of the National Center for Biotechnology Information. "Now we are starting to see the true complexity of life."

Despite the quagmire in which their present efforts have landed them, biologists have not in any way despaired of confirming the conventional thesis, that life evolved on Earth from natural chemical processes. But a ferment of rethinking and regrouping is under way.

Up until now, searchers in the treasure hunt for the universal ancestor have followed a hallowed chart known as the ribosomal RNA phylogenetic tree. This is a family tree drawn up by Dr. Woese and based on a gene used by all living cells to specify ribosomal RNA, or ribonucleic acid, a component of the machinery that translates genetic information into working parts. It was this tree that led Dr. Woese to recognize the tripartite division of living things and that one of the three kingdoms belonged to the Archaea, previously assumed to be a weird sort of bacteria.

Many of the deepest branches in Dr. Woese's tree, those that join nearest to the three-way junction of the kingdoms, turned out to belong to organisms that live at high temperatures, like those in the fuming springs in Yellowstone Park or in the volcanic vents that gash the ocean floor. That clue fit in well with new theories that life originated at volcano-like temperatures.

With the new ability to decode the full DNA of a microbe, it is these high-temperature microbes that biologists have chosen for some of their first targets. *Aquifex aeolicus,* a denizen of Yellowstone Park that lives at a temperature just five degrees less than the boiling point of water, is the deepest branching of all known bacteria. In the light of evidence suggesting that the oldest region of the ribosomal RNA tree lies on the branch leading into the bacterial kingdom, *Aquifex* had grounds for claiming to be the nearest living cousin of the universal ancestor.

But the sequence of the *Aquifex* genome, reported in the journal *Nature,* has yielded only disappointment. For one thing, the microbe appears to have only one gene, called a reverse gyrase, that is not found in organisms that live at ordinary temperatures. That suggests that it may be quite easy for microbes to switch between high and normal temperatures, said Dr. Ronald V. Swanson, a member of the *Aquifex* team who works at the Diversa Corporation of San Diego.

A second blow is that with the full genome sequence in hand for *Aquifex* and a dozen other microbes, biologists can draw up family trees based on other genes, besides the ribosomal RNA gene that provided the original map. And the trees based on other genes show different maps that do not agree with the ribosomal RNA map. "Each picture is different, so there is tremendous confusion," Dr. Woese said.

A basic source of the confusion is that in the course of evolution whole suites of genes have apparently been transferred sideways among the major branches. Among animals, genes are passed vertically from parent to child, but single-cell creatures tend to engulf each other and occasionally amalgamate into a corporate genetic entity. It has long been argued that mitochondria, the tiny organelles that handle the energy metabolism of eukaryotic cells, were once free-living bacteria that were enslaved by an early eukaryote. Mitochondria still possess their own, bacteria-like DNA, but many of their genes have emigrated into the eukaryotic cell's own DNA in the nucleus.

Horizontal transfer of genes between kingdoms would severely tangle up the lines in family trees. "What impresses me is that the pattern of genes we see among organisms is not reduced to total chaos," said another member of the *Aquifex* team, Gary J. Olsen of the University of Illinois.

Presumably because of sideways gene traffic in the distant past, both Archaea and Eukarya seem to rely on bacterial-type genes to manage much of their general chemical metabolism. (The Eukarya, thought to be descended from the Archaea, rely on archaean-type genes to manage their DNA and to translate its genetic information into protein products.)

"It's possible that bacterial genes have swept all over the world and replaced everything else that existed, so some of the features of the last common ancestor may have been erased from the face of the planet," Dr. Koonin said.

But no one is abandoning the search for the universal ancestor. "My biggest fear is that evolution would be indecipherable because of all the random changes that took place," said Craig Venter of the Institute for Genomic Research in Rockville, Maryland. "The good news is that that is clearly not the case. I think it will be completely decipherable, but because of horizontal transfer, the tree may look more like a neural network," he said, referring to the crisscross pattern of a neural computing circuit. Dr. Venter, who pioneered the sequencing of microbial genomes, estimated that 50 to 100 more genomes would have to be sequenced to help triangulate back to the last common ancestor.

Evolutionary biologists are working on several approaches for seeing beyond the confusion caused by lateral transfer. Computational biologists like Dr. Koonin believe that it is already possible to identify 100 or so genes that the common ancestor must have possessed, mostly ones that manage DNA and its translation into proteins, and that others can be added with varying degrees of certainty.

Most biologists still favor the standard view that the universal ancestor, already a quite sophisticated organism that had come a long way since the origin of life, first branched into the Prokarya (bacteria) and the Archaea. Later the Eukarya branched off from the Archaea, but accepted many genes from the Prokarya. Dr. Koonin describes the eukaryotic cell as a "palimpsest of fusions and gene exchanges," referring to a manuscript that has been written over with new text.

But some important eukaryotic genes have no obvious predecessors in either the archaean or bacterial lines. The family of genes that makes the stiff framework of eukaryotic cells, known as the cytoskeleton, seems to appear out of nowhere. "The absence of sequences closely related to the slowly changing proteins of the eukaryotic cytoskeleton remains unsettling," Dr. Russell F. Doolittle of the University of California, San Diego, wrote in the journal *Nature*.

Another evolutionary biologist, Dr. W. Ford Doolittle of Dalhousie University in Halifax, Nova Scotia, has an explanation, though one he concedes does not yet enjoy the company of evidence. He argues that there might have been many lost branches of the tree of life before the universal ancestor. One branch, a fourth kingdom of life, might have contributed the cytoskeleton genes to the Eukarya before becoming extinct.

A new and far-reaching theory about the universal ancestor has been developed by Dr. Woese. His colleagues said the theory envisages all three kingdoms emerging independently from a common pool of genes. The pool was formed by a community of cells that frequently exchanged genes by lateral transfer. The price of membership in the community was the use of the same genetic code, according to Dr. Woese's theory, which is how the code came to be almost universal.

The community of protogenomes quickly shared new innovations among themselves, in Dr. Woese's view, and the system evolved by producing more complicated proteins, the working parts of the cell. The genetic code was at first translated rather inaccurately, so the proteins it produced were short and limited in capability. But the code became more accurate, and the proteins more complex, driven by the advantage that more capable proteins conferred.

At a certain stage of complexity, design decisions may have limited cells' ability to exchange genes, and the ancestral pool would have split into the three kingdoms seen today, the theory suggests.

It is possible, of course, that evolution's early traces have become too faint to decipher. And at the back of researchers' minds is another worry that makes them throw up their hands because it cannot be addressed scientifically: that life might have arrived on Earth from elsewhere.

Life seems to have popped up on Earth with surprising speed. The planet is generally thought to have become habitable only some 3.85 billion years ago, after the oceans stopped boiling from titanic asteroid impacts. Yet by 3.5 billion years ago, according to the earliest fossil records, living cells were flourishing, and there are indirect signs of life even earlier, in rocks that are 3.8 billion years old.

"There's the 'gee whiz' point of view—how can life possibly have evolved in three hundred million years—which I think is still a problem," said Dr. Doolittle of Halifax. But life arriving from outer space is a theory, he said, that "leaves you stunned—there is nothing more you can say after that."

This narrowing window of time may be less embarrassing than it seems. Biologists are warming to the view that the emergence of life from chemical precursors is a quite probable event that does not require billions of years to get under way. "You put a selective hammer on it and it happens fast," said Norman R. Pace, an evolutionary biologist at the University of California, Berkeley, referring to the force of natural selection. "It's shockingly fast, maybe just tens of millions of years."

Still, many more years of evolution presumably passed before the universal ancestor, a quite sophisticated genetic system, attained its final form.

If the ancestor was a pool of organisms as Dr. Woese suggests, and not a definable species, it may be even harder to capture its likeness. But knowledge about this distant era at the dawn of life is moving so fast that few biologists are troubled by setbacks like the *Aquifex* dead end or the discordant family trees. "I'm unwilling to say we'll never know about anything because we have come so far in the last two decades," Dr. Pace said.

Some family tree problems, after all, have exact solutions. For example, Dr. Doolittle of Halifax wrote a comment on the article by Dr. Doolittle of San Diego that they had discovered the reason for their common name: They shared a common ancestor eight generations back.

—Nicholas Wade, April 1998

2

DISCOVERING
HUMAN GENES

Genes shape the human body and mind. They do not determine everything about a person, but they set the framework of what is possible. Legs here, arms there, no wings, shell or sting. Many diseases are due to malfunctions in particular genes. Some of the malfunctions are inherited, some arise through damage to the genes in the course of a person's lifetime.

Doctors, therefore, as well as biologists, are intensely interested in discovering all the human genes, in the hope that means can be found of correcting genetic errors.

The hunt for human genes is taking place at two levels. One, the bottom-up approach, is to study the pedigree of families in whose members a disease is prevalent. By looking for genetic markers—known regions of DNA—that are always inherited along with the disease, researchers can sometimes locate the causative gene, which will lie reasonably close to the marker. The genes that in defective form cause Huntington's disease and the inherited forms of breast cancer were located in this way.

The top-down approach is more ambitious: Its goal is to locate and identify every human gene by sequencing all three billion units of human DNA. The Human Genome Project, as it is known, is scheduled to be completed by the year 2005.

Determining the sequence of the chemical letters in DNA is not an end in itself, however, but the first step to a larger goal. Biologists must first detect where the genes lie on the DNA, and discover what each gene does.

One important way of interpreting human genes is by relating them to genes already discovered in other animals, like the *Drosophila* fruit fly, the *Caenorhabditis elegans* roundworm and the mouse. These little animals have been intensively studied in laboratories for many years, and can be genetically manipulated in ways that would be impossible in humans. Because of its relevance to humans, work on sequencing the worm's genome is funded by the human genome program at the National Institutes of Health. A large community of biologists now studies the worm, and their findings will be of increasing help in understanding the operating principles of the human genome.

Many genes from the worm and other organisms have been analyzed and their sequences deposited in the computer data banks. These data banks, like the GenBank operated by the National Institutes of Health in Bethesda, Maryland, have become an important resource for biologists. When the DNA sequence of an unknown gene is found, biologists can search the data banks for genes in other species that have a similar DNA structure and, hopefully, a known function. The unknown gene is likely to perform a similar role.

The articles that follow describe the efforts to understand DNA and the genomic level, as well as the work being done in parallel to find and understand individual genes.

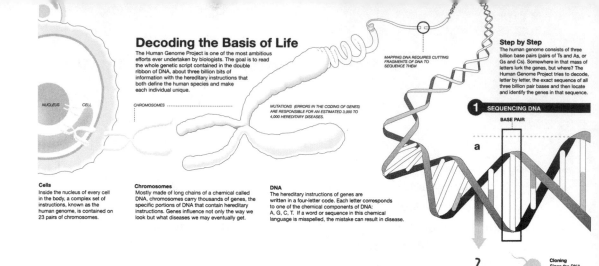

Decoding the Basis of Life

The Human Genome Project is one of the most ambitious efforts ever undertaken by biologists. The goal is to read the whole genetic script contained in the double ribbon of DNA, about three billion bits of information with the hereditary instructions that both define the human species and make each individual unique.

NUCLEUS CELL

CHROMOSOMES

MAPPING DNA REQUIRES CUTTING FRAGMENTS OF DNA TO SEQUENCE THEM

MUTATIONS (ERRORS IN THE CODING OF GENES) ARE RESPONSIBLE FOR AN ESTIMATED 3,000 TO 4,000 HEREDITARY DISEASES.

Step by Step
The human genome consists of three billion base pairs (pairs of Ts and As, or Gs and Cs). Somewhere in that mass of letters lurk the genes, but where? The Human Genome Project tries to decode, letter by letter, the exact sequence of all three billion base pairs and then locate and identify the genes in that sequence.

1 SEQUENCING DNA

BASE PAIR

a

Cells
Inside the nucleus of every cell in the body, a complex set of instructions, known as the human genome, is contained on 23 pairs of chromosomes.

Chromosomes
Mostly made of long chains of a chemical called DNA, chromosomes carry thousands of genes, the specific portions of DNA that contain hereditary instructions. Genes influence not only the way we look but what diseases we may eventually get.

DNA
The hereditary instructions of genes are written in a four-letter code. Each letter corresponds to one of the chemical components of DNA: A, G, C, T. If a word or sequence in this chemical language is misspelled, the mistake can result in disease.

Cloning
Since the DNA chain is too long, mappers cut it into fragments (about 150,000 base pairs long). In this example, each of three fragments is multiplied by being grown in bacteria, a process called cloning.

BACTERIA
CLONES
150,000 base pairs

b

Mapping
The cloned fragments of DNA are created so that they overlap. The clones are then put in order so as to reconstruct the original sequence.

150,000 bp

c

Reducing
The mapped DNA fragments are broken down by sound into random length pieces, which are again cloned. The pieces can be read now by DNA sequencing machines.

1,500 bp

d

DNA STRANDS SEPARATED

C T A G

C T A G A G C T

CTAGAGCT

500 bp

Sequencing
The two strands of DNA are separated and analyzed individually. The four bases of DNA (Ts, As, Cs, Gs) are dyed with four different colors and then separated into a ladder-like pattern through chemical reactions. A computer program called Phred reads the pattern and gives the order of the bases in the DNA sequence.

e

PHRAP

Reassembling
Another program (Phrap) orders the fragments. Longer sequences are reassembled to read the complete DNA sequence of the chromosomes.

The Struggle to Decipher Human Genes

IN A DRAB GRAY BUILDING in the industrial outskirts of St. Louis, a team of some 200 people is working 19 hours a day in pursuit of the ultimate self-knowledge. They are spearheading the effort to sequence the human genome by 2005.

The odds for success at this point are not overwhelming. At the end of this month, the project will be halfway through its planned 15-year course, yet only 3 percent of the genome has been completed. Of the nine American centers involved in the pursuit, just one, the Genome Sequencing Center here in St. Louis, is deciphering human DNA at a significant rate. Its only peer is the Sanger Center in Hinxton, England. Together the two cen-

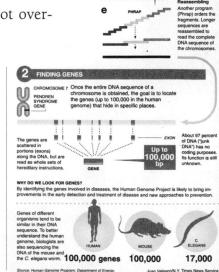

2 FINDING GENES

CHROMOSOME 7
PENDREN SYNDROME GENE

Once the entire DNA sequence of a chromosome is obtained, the goal is to locate the genes (up to 100,000 in the human genome) that hide in specific places.

The genes are scattered in portions (exons) along the DNA, but are read as whole sets of hereditary instructions.

EXON

Up to 100,000 bp

GENE

About 97 percent of DNA ("junk DNA") has no coding purposes. Its function is still unknown.

WHY DO WE LOOK FOR GENES?
By identifying the genes involved in diseases, the Human Genome Project is likely to bring improvements in the early detection and treatment of disease and new approaches to prevention.

Genes of different organisms tend to be similar in their DNA sequence. To better understand the human genome, biologists are also sequencing the DNA of the mouse and the C. elegans worm.

HUMAN 100,000 genes
MOUSE 100,000
C. ELEGANS 17,000

Source: Human Genome Program, Department of Energy.   Juan Velasco/N.Y. Times News Service

ters have produced half of the 106 million letters of human DNA so far completed.

The goal of sequencing the entire three billion letters of human DNA is not just technically ambitious. The coiled double ribbon of DNA holds the genetic instructions to make and operate the human organism. It bears the record of how humans differ from apes, the saga of early human migrations and the programming variations that help make each individual unique. The genome is the basis for much of what scientists can hope to explain about the physical aspects of human life.

"For the first time, we humans are reducing ourselves down to DNA sequences," said Dr. Robert Weinberg, a leading cancer biologist at the Whitehead Institute in Cambridge. "We're not talking about how butterflies fly or trees grow: We are dealing here with the mystery of the human spirit. Analysis of these sequences will not define the essence of humanity, but aspects of human beings that have hitherto been as awe-inspiring will be reduced to rather banal biochemical explanations, and that's not altogether heartening—maybe the mystery is good."

As might be expected in so ambitious an undertaking, opinions differ as to whether the full human sequence can be completed on schedule. Dr. Francis S. Collins, the director of the National Human Genome Research Institute, said he was "quite optimistic this goal will be achieved, and without depending on a bolt-from-the-blue breakthrough." But the costs of sequencing are not coming down as fast as hoped, and the scientists involved in the effort are concerned that the necessary resources may not be available. One of them, Dr. Craig Venter of the Institute for Genomic Research in Rockville, Maryland, said that "every single group has fallen behind, even us."

The genome project has always seemed a stretch, and it still depends on an invent-as-you-go approach. For no other animal, not even the simplest, has the full genetic programming yet been decoded. Only recently have the first bacteria, with genomes of a mere one million or so letters of DNA, yielded to sequencing, the term for working out the order of the chemical units in DNA. The human genome is 3,000 times as large.

Sequencing the human genome is a tangible goal but only part of a much larger endeavor, that of understanding what the genetic instructions in the genome mean. Postgenomics, as this interpretive process is sometimes called, has already begun in earnest. A new discipline known as bioinformatics, or computational biology, has sprung up to handle and interpret the streams of DNA now entering computer databases. GenBank, the DNA database run by the National Center for Biotechnology Information, already holds more than a billion DNA bases from human and other species, and has grown so popular that some 8,000 biologists consult it daily.

Another spur to postgenomics has been a technique for fishing out small snippets of genes from the genome. The DNA sequence of these fragments can be programmed into a new device called a DNA chip, enabling the chip to tell which genes are expressed, or switched on, in particular tissues of the body. Biologists can now hope to understand many diseases at the level of the human cell by comparing gene expression in normal and diseased tissue.

"The fruits of the genome project will enormously speed our efforts to understand human diseases, both inherited and those that strike in lifetime," Dr. Weinberg said.

So vigorous is the thrust into postgenomics that those responsible for sequencing the DNA fear they may not get the support to complete their task on schedule. "You can't divert resources to other things without suffering on the sequencing," said Dr. Robert H. Waterston, director of the Genome Sequencing Center, which is part of the Washington University School of Medicine in St. Louis.

Dr. Waterston presides over a remarkable enterprise, which is part research laboratory and part industrial process. His colleagues, as arcane a group of specialists as one could find, are known as mappers, sequencers, finishers and annotators.

The mappers take fragments of human DNA and try to map them to the exact position on the chromosome from which they were derived. The 23 pairs of human chromosomes are the units in which the DNA is packaged, and the St. Louis center is currently focusing on chromosome 7, which is 170 million DNA letters, or base pairs, long.

The mapped fragments, each about 150,000 base pairs of DNA in length, are handed over to the four teams of sequencers, who break them down into shorter pieces some 1,500 bases long. These are fed into the center's 70 Applied Biosystems sequencing machines. In a clever process worked out by Frederick Sanger of Britain, the machines tag the DNA with four different dyes, one for each of the four letters in the DNA alphabet. The order in which the colors stream out of the machines represents the order of the bases in the DNA.

The finishers' job is to arrange the short sequences of DNA so that they overlap, allowing that of the parent fragments to be reconstructed. The problem is that in many cases the sequences do not overlap because the DNA has evaded one or another of the steps in the production process. The finishers have ways of bridging the gaps, some of which have been automated, but closing the gaps remains one of the major technical obstacles in the genome project.

The finishers rely heavily on two computer programs written by Philip Green of the University of Washington in Seattle. One program, named Phred, scans the output of the sequencing machines and calls the order of the bases along with the level of confidence that can be placed in each call.

The other, Phrap, tries to assemble the emerging sequences into overlapping sections. Much of the task is like the mental torture of doing a jigsaw puzzle in which all the pieces are virtually identical. The reason is that some 97 percent of human DNA consists of a variety of identical sequences repeated over and over. The repetitive sequences have no known function and used to be called junk DNA.

Perhaps foreseeing limits on Congress's interest in paying to sequence junk, biologists now refer to these DNA wastelands more delicately as "noncoding DNA," meaning DNA that does not code for genes. Only 3 percent of the human genome specifies working genes, of which there are thought to be between 60,000 and 100,000.

The world's biologists are so eager to get their hands on the data that new DNA sequences are posted every night on the center's Web site. But months' more work may remain as the raw data are checked for accuracy, the present standard being no more than one error per 10,000 bases, and analyzed by the annotators.

The annotators' job, the culmination of the whole process, is to locate and identify genes. Picking out human genes from a DNA sequence is not easy: There is no punctuation, no known "start here" signal, just eye-glazing rows of As, Gs, Ts and Cs—the four letters of the DNA alphabet. The full genome, when and if ever printed out, would take up 200 volumes the size of telephone directories of 1,000 pages each. More probably it will just be recorded on a CD-ROM.

Analysis of the human genetic program is entrusted to computer programs as much as possible. The computer marks the sites of probable genes and, where possible, assigns the genes a function by comparing their DNA sequences with those of genes of known function from other species. Finally, large chunks of annotated DNA sequence are submitted to GenBank.

A foretaste of the fruits to be expected from sequencing the human genome is emerging from that of the *C. elegans* roundworm, whose genome is 100 million base pairs long, about the size of a single human chromosome. The worm's genome is being sequenced at a cost of $40 million at the Genome Sequencing Center and at the Sanger Center in Britain. The effort, which is almost complete, served as a test run for the techniques being used with the human genome. Early success with the worm helped Dr. Waterston and his English colleague, Dr. John E. Sulston, to persuade skeptical colleagues in 1993 that they were ready to move from the mapping to the sequencing phase of the human genome.

The worm's genome is of enormous interest to the biologists who are studying the organism, and their findings will help interpret the human genome, because the two organisms, despite their evolutionary distance, have genes of similar DNA sequence.

To help understand the *C. elegans* genome, the St. Louis center has been sequencing the genome of another species of roundworm, known as *C. briggsae*. The genes of the two worms turn out to be very similar, but the noncoding DNA between the genes is entirely different. The pattern demonstrates the power of natural selection, the motivational force of evolution: The noncoding regions of DNA are evidently free to mutate without constraint, whereas the DNA that codes for genes must stay more or less constant so as to avoid generating a misshapen protein that will kill the organism.

Is the Human Genome Project going well? To judge by all the postgenomic activity, and the enormous recent investment in genomics by the pharmaceutical industry, the project is already a startling success. But will it be finished on time?

At a recent meeting of the genome project's main participants last month in Bermuda, a chart was displayed comparing how much DNA each center had promised to sequence at last year's meeting and how much it had in fact done. "It was a useful reality check," Dr. Collins said, suggesting that some of the promises might have been efforts to impress financing agencies.

The Human Genome Project is biologists' first serious foray into Big Science, an endeavor with which physicists have long been familiar in the form of constructing particle accelerators. One physicist, Dr. Steven E. Koonin, vice president of the California Institute of Technology, headed a group of physicists that criticized the Human Genome Project recently for its lack of central coordination. "The managerial structure is much looser than in a physics project," he said.

Biologists, who take pride in the project's very lack of central control, do not respond to the physicists' opinions with conspicuous deference. But Dr. Koonin's concerns about the lack of good comparative cost data from the different sequencing centers are echoed by an internal critic. "I have been calling the group of scientists involved in human sequencing the Liars' Club," said Dr. Venter, a biologist of independent views who has his own institute. "They all have a different way of calculating their costs and the amount of sequencing they have actually accomplished."

Comparative cost is an issue the sequencers at the Bermuda meeting agreed should now be seriously addressed. Dr. Waterston said his center had brought its operating costs down to 40 cents per DNA base. But even at that price, completing the genome would cost $1.2 billion.

Sequencing the human genome would be cheap at almost any price, Dr. Venter noted, given that the search for the Huntington's disease gene alone cost more than $100 million. But he doubted if costs could fall much more because competition is driving up wages.

Dr. Venter also believes that the mapping problem has not yet been solved. "A billion was spent to map the genome, but out of all the upfront money spent for mapping, there are no sequence-ready clones," Dr. Venter said, referring to the essential first step of acquiring DNA fragments of known position on the chromosomes.

Dr. Collins, the National Institutes of Health's director for the genome project, said that there was "no major technical problem" in mapping the fragments and that Dr. Venter, who does not do his own mapping, was probably worried about keeping his sequencing machines busy.

Dr. Collins said that getting the human DNA sequence completed was a higher priority for his institute than the interpretive projects. "We are not going to take our eyes off the ball," he said in reference to Dr. Waterston's concerns about a diversion of effort.

The *C. elegans* genome, which served as the pilot project for that of the human genome, is nearing completion but the hardest parts of the sequence, like closing the remaining gaps, have been left until last.

"I doubt the worm will be finished this decade if they are held to the standard of closing all the gaps," Dr. Venter said.

Dr. Waterston, however, said that the worm's genome would be completed by the end of this year and that the remaining problems were manageable.

Dr. David Botstein, a leading biologist at Stanford University who has helped shape the Human Genome Project, said that in his view the enterprise was reasonably well on track, despite having received about half the money originally envisioned.

The National Academy of Sciences committee that gave the project its imprimatur "assumed that virtually all the sequencing would be done during the last half of the project, so there's no cause for alarm at that end," Dr. Botstein said. "The cost of sequencing has not fallen as much as hoped," he said, but he added that it "doesn't have to fall much further to be affordable and doable in the time required."

"It seems almost a miracle to me," Dr. James Watson, the codiscoverer of DNA, wrote recently, "that fifty years ago we could have been so ignorant of the genetic material and now can imagine that we will have the complete genetic blueprint of man." Dr. Watson said he was not disheartened by the state of progress reported at the Bermuda meeting, even though he had not foreseen that "everyone would regard the genome as gold," making it hard for the sequencers to prevent trained people from being hired away by pharmaceutical companies.

Sequencing efficiently is still difficult but "St. Louis does it beautifully, so it can be done," Dr. Watson said.

—NICHOLAS WADE, March 1998

Dainty Worm Tells Secrets
of the Human Genetic Code

IN JUNE 1997 nearly a thousand biologists met in Madison, Wisconsin, to discuss the affairs of a speck of protoplasm with a grandiloquent name, a barely visible, almost transparent worm called *Caenorhabditis elegans*. The worm holds the same secrets of life as other animals, but may be the first to yield them, an event that would deeply influence biology and medicine.

The tiny creature at the focus of this scrutiny has several surprising properties. One is its daintiness, recognizable even by those who never expected to see any redeeming qualities in a worm. Under the microscope, *C. elegans* sweeps its head gracefully from side to side, the crystalline body follows in a sinuous wave, and the viewer can see at once how the little worm invited the *"elegans"* part of its name.

Another surprise has been the closeness of its genetic kinship to humans. Most of the human genes being discovered turn out to have counterpart genes in the worm, ones so similar in chemical structure that they must have evolved from the same parent DNA in the distant common ancestor of both worms and humans. Even after all these eons, the closeness is real enough that in several cases biologists have been able to insert the human version of a gene in place of the worm's own copy. *C. elegans* gets along just fine with its human replacement part.

C. elegans is a kind of worm known as a nematode and is no relation to its fellow soil-dweller the earthworm. It owes its escape from obscurity to having been chosen along with the *Drosophila* fruit fly, the zebra fish and the mouse, as a model animal, one that biologists study in depth for the light it may shed on general principles of life.

The people who study the worm are an unusually close-knit fraternity, almost an extended family, because most of the principal scientists in the

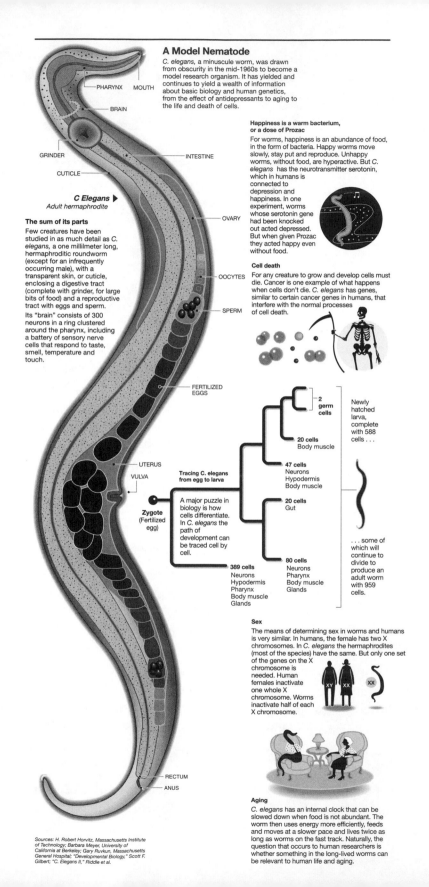

A Model Nematode

C. elegans, a minuscule worm, was drawn from obscurity in the mid-1960s to become a model research organism. It has yielded and continues to yield a wealth of information about basic biology and human genetics, from the effect of antidepressants to aging to the life and death of cells.

PHARYNX MOUTH

BRAIN

GRINDER

CUTICLE

INTESTINE

C Elegans ▶
Adult hermaphrodite

OVARY

OOCYTES

SPERM

The sum of its parts

Few creatures have been studied in as much detail as *C. elegans*, a one millimeter long, hermaphroditic roundworm (except for an infrequently occurring male), with a transparent skin, or cuticle, enclosing a digestive tract (complete with grinder, for large bits of food) and a reproductive tract with eggs and sperm.

Its "brain" consists of 300 neurons in a ring clustered around the pharynx, including a battery of sensory nerve cells that respond to taste, smell, temperature and touch.

FERTILIZED EGGS

UTERUS

VULVA

RECTUM

ANUS

Happiness is a warm bacterium, or a dose of Prozac

For worms, happiness is an abundance of food, in the form of bacteria. Happy worms move slowly, stay put and reproduce. Unhappy worms, without food, are hyperactive. But *C. elegans* has the neurotransmitter serotonin, which in humans is connected to depression and happiness. In one experiment, worms whose serotonin gene had been knocked out acted depressed. But when given Prozac they acted happy even without food.

Cell death

For any creature to grow and develop cells must die. Cancer is one example of what happens when cells don't die. *C. elegans* has genes, similar to certain cancer genes in humans, that interfere with the normal processes of cell death.

Tracing C. elegans from egg to larva

A major puzzle in biology is how cells differentiate. In *C. elegans* the path of development can be traced cell by cell.

Zygote
(Fertilized egg)

- 2 germ cells
- 20 cells
  Body muscle
- 47 cells
  Neurons
  Hypodermis
  Body muscle
- 20 cells
  Gut
- 80 cells
  Neurons
  Pharynx
  Body muscle
  Glands
- 389 cells
  Neurons
  Hypodermis
  Pharynx
  Body muscle
  Glands

Newly hatched larva, complete with 588 cells . . .

. . . some of which will continue to divide to produce an adult worm with 959 cells.

Sex

The means of determining sex in worms and humans is very similar. In humans, the female has two X chromosomes. In *C. elegans* the hermaphrodites (most of the species) have the same. But only one set of the genes on the X chromosome is needed. Human females inactivate one whole X chromosome. Worms inactivate half of each X chromosome.

XY XX XX

Aging

C. elegans has an internal clock that can be slowed down when food is not abundant. The worm then uses energy more efficiently, feeds and moves at a slower pace and lives twice as long as worms on the fast track. Naturally, the question that occurs to human researchers is whether something in the long-lived worms can be relevant to human life and aging.

Sources: H. Robert Horvitz, Massachusetts Institute of Technology; Barbara Meyer, University of California at Berkeley; Gary Ruvkun, Massachusetts General Hospital; "Developmental Biology," Scott F. Gilbert; "C. Elegans II," Riddle et al.

field underwent their apprenticeship in the same laboratory in Cambridge, England. Worm people differ from other tribes of biologists, like those who study the fly. They have their own traditions, including a more holistic approach to their animal, and a barn-raising ethos of creating tools that their whole research community can use.

The output of this worldwide effort is something like a manual on how the worm works, though only the very first chapters have been written. Even a partial manual is of enormous help in trying to understand a system that has some 17,000 instructions (its genes) and a larger number of different working parts (its proteins). The worm's nervous system has only 302 neurons, but there are 118 different types with some 5,000 interconnections, making for a complexity that has so far defied analysis.

Humans, by contrast, have some 65,000 genes. Understanding how so many genes interact will not be easy, which makes many biologists eager to try to figure out the worm first. "It's a little gold mine," said Dr. Cynthia Kenyon of the University of California at San Francisco. "Basically, it's an animal that does everything animals do, except arithmetic. It's like a little streamlined helper for these bigger, messier animals."

As a model animal for molecular biologists, *C. elegans* is the single-handed creation of Dr. Sydney Brenner, one of the pioneers in working out the genetic code. Looking around for a challenge in the early 1960s, Dr. Brenner decided that the brain deserved attention but that the fruit fly's, with its 100,000 neurons, was too difficult. *C. elegans's* seemed suitably limited. The worm had other attractive features, like its transparency, and ease of culture.

Dr. Brenner's project got under way in 1965 with support from the Medical Research Council, the British counterpart of the National Institutes of Health. Starting a model animal from scratch, especially when so much groundwork had already been laid with the fruit fly, was not an obvious choice. "Because of his charisma and personality, he sold the idea to the MRC," said Dr. John White, one of Dr. Brenner's original Cambridge group and now at the University of Wisconsin at Madison.

It took Dr. Brenner nine years to explore the worm's genetics and prove that useful biology could be done with it. The handful of English scientists he recruited to the project were joined by a succession of American postdoctoral students, who returned to the United States and set up their own laboratories.

At the first scientific meeting devoted to *C. elegans,* held at Woods Hole, Massachusetts, in 1977, 70 biologists turned up. At the June 1997 gathering, there were 974, almost one for each cell in the adult worm. Some 200 laboratories around the world now study the little nematode, in the estimate of Dr. Robert K. Herman, who runs the world's worm stock center at the University of Minnesota. The center mailed out 189 shipments of worms in 1980, 4,004 in 1996. Fortunately, the worms can be frozen until needed. All of the center's 2,700 strains fit into eight large freezers.

No one could confuse the *C. elegans* community with the group that studies the *Drosophila* fruit fly. When fly people find a new gene, they give it a colorful name, like Tudor or rutabaga. Worm people name all their genes with a three-letter word and a number. The Groucho gene in *Drosophila* becomes unc-37 in worm-speak, suggesting why some fly people claim worm people have no sense of humor.

But worm people have traditions more appealing than their system of gene nomenclature. Materials and information are freely shared. "It is the kind of community that has maintained the standards of scientific practice that everybody wants and hardly anybody achieves," Dr. Brenner said.

Talks at worm meetings are given by the young graduate students who did the experiments, while the laboratory chiefs stay out of the spotlight. Everyone listens to all the talks, whereas fly meetings are broken up into parallel sessions.

Sense of humor or not, worm people do occasionally let their meetings get completely of hand. After a food fight one year, irritated hosts presented a $2,000 bill for cleaning the drapes.

Asked the purpose of the *C. elegans* project, worm biologists respond from a variety of perspectives. Some see the goal as what would be an extraordinary exercise in reductionism, the explanation of an animal's total behavior in terms of the atoms and molecules of which it is built. Others see the worm as a step toward understanding problems in biology, not as an end in itself.

The purpose, Dr. Herman said, "is to understand how the worm is made and how it works, in terms of its molecules." Dr. Robert H. Waterston of Washington University School of Medicine in St. Louis said, "The idea is to understand every behavior in terms of the genes that produce it." But Dr.

Martin Chalfie of Columbia University sees the worm as a means to address such interesting problems in biology as how 17,000 genes work together.

One prized tool for understanding *C. elegans* is its parts list. The animal is made up of exactly 959 cells, each with a name. The family tree whereby cells of the adult worm develop from the single cell of the fertilized egg was worked out largely by Dr. John E. Sulston of the Sanger Center in Cambridge and Dr. H. Robert Horvitz of the Massachusetts Institute of Technology. Although the work was completed 14 years ago, there is still nothing similar for any other model animal.

Another development that has spread *C. elegans*'s fame beyond the world of worm people is the project to sequence its genome, which means determining the order of all the chemical letters in its DNA. Only 1 percent of the three billion letters of human DNA have been sequenced, and the completion date is not until 2005. The worm's genome, by contrast, is nearly done.

When medical biologists track down a new human disease gene, they often check the data banks and find a related worm gene, prompting proposals of collaboration with worm biologists to see how the gene works. Worms do not live long enough to suffer the maladies of age. Still, even the human genes that in defective form cause diseases like Alzheimer's and colon cancer have counterpart genes in *C. elegans*.

Of the 100 million letters in the worm's DNA, 67 million have already been decoded and analyzed, and another large chunk is nearly ready. The two chief architects of the decryption process, Dr. Waterston, of Washington University, and Dr. Sulston, of the Sanger Center, expect to wrap up their work by the end of next year, a landmark that will make *C. elegans* the first animal genome to be sequenced.

The full DNA sequence of *C. elegans* will give biologists their first glimpse of the genetic programming required to make and operate an animal. One surprise is the enormous number of genes indicated by the sequencing available so far. Using the standard method of knocking out genes with mutation-causing chemicals, biologists have discovered only a few thousand genes. The first encyclopedia on the worm, published in 1988, estimated that *C. elegans* possessed 4,000 genes, less than a quarter of the number now expected by the DNA sequencers. The emergence of these extra genes is proof of how much remains to be learned about the worm and how complicated living organisms are.

Worm researchers have already made several discoveries that have made other biologists take note. One is the phenomenon of cell death. In tracing the cell lineage tree of *C. elegans,* Dr. Horvitz and Dr. Sulston found that certain cells were always destroyed during development, and Dr. Horvitz went on to find that the destruction was programmed by the cells' own genes. It is now known that almost all animal cells have a self-destruct mechanism, and that in humans the mechanism is one of the body's leading defenses against cells that become genetically unstable and cancerous.

Another prominent field of research that started in *C. elegans* is that of the axon guidance factors, known as netrins, which tell nerve cells where to send their communication fibers and help wire up an animal's nervous system. The gene that makes the factor was first found by a worm researcher, Dr. Edward M. Hedgecock of Johns Hopkins University.

Worm biologists have also made important contributions to the study of cell signaling, the elaborate chain of reactions that translates a message received at the surface of a cell into the switching on or off of genes inside the cell's nucleus. Many worm labs have pored over the details of how a little group of 22 cells in the embryo exchange signals and organize themselves into forming a tube between the worm's skin and its uterus. The problem, as Dr. White puts it, is that of how to do plumbing without a plumber.

The vulva of a tiny worm might seem of little interest to anyone but worm biologists and male worms, but the cell signaling mechanisms involved in organizing the vulval cells turn out to be the same as those used in certain human cells. In humans, mutations in the genes that program the component parts of these signaling pathways can lead to cancer.

Dr. Brenner's gamble on the worm has paid off richly, even though it has not developed in quite the way he expected. Many worm people "consider that the initial goal of using the worm to understand the nervous system has so far been a dead end," said Dr. Rachel Ankeny, a historian at the University of Pittsburgh who is writing an account of the worm project. The worm's neural anatomy was successfully reconstructed over 14 years by Dr. White from 20,000 serial cross-sections.

But the wiring diagram, with its 5,000 interconnections, has not yet yielded many answers about how the worm behaves. "It is actually very complicated," Dr. Horvitz said. "People thought naively we could look at the nervous system and derive behavior from it, which is completely untrue."

"We got the wiring diagram but the second problem, computing behavior, never really took off," Dr. Brenner said. "My own view is that it was a problem for younger people."

As the sequencing of the nematode genome nears completion, worm people have a sense of confidence that their field is pushing the frontiers of biology, leading the way toward solutions for many problems that will be important in human biology and medicine. "The problems of *C. elegans* are to a great extent the problems of biology, many of which are unsolved, or solved at one level of explanation, Dr. Horvitz said. "So there is a vast wealth of biological knowledge that can be extracted from the worm."

Asked why he no longer worked on *C. elegans,* Dr. Brenner explained his changes of field, from molecular biology in the 1960s and more recently from the worm, by referring to J. D. Bernal, the English crystallographer and historian of science: "Bernal said that science is like a chess game and that it is given to very few of us to be allowed to play the end game. When you can't play the end game, the middle game is extremely boring, so the best thing is to play the opening game, and I think I'm pretty good at it and I enjoy it. So many fields are possessed by the people that start them, but I don't feel that. I feel it was great fun bringing this animal out of nature and into the lab, and I will go and do something else."

—NICHOLAS WADE, June 1997

A Bold Shortcut to Human Genes

IN A FORMER CERAMICS FACTORY on the outskirts of the nation's capital, Dr. J. Craig Venter has created a biologist's dream: a high-technology laboratory dedicated to the sequencing of DNA. Out of it is flowing a rich torrent of new information, chemical sequences from many of the 75,000 or so genes that specify a human.

"I don't see how this cannot revolutionize biology," Dr. Venter said. His delight is in generating and analyzing the raw text in which the story of human evolution is written. But this affable, soft-spoken man harbors a barely veiled ambition to slay giants. He intends to execute a quick end run around the U.S. government's $3 billion Human Genome Project.

Washington's goal is to spell out each of the three billion chemical letters in the human gene set. Dr. Venter is focusing on the tiny portion of DNA that harbors the genes, while ignoring the vast stretches that have no known purpose.

Other biologists do not doubt that he can generate a lot of sequence data—the endless strings of As, Ts, Gs and Cs in which the genetic code is written—but they are waiting to see what sense he can make of it. Dr. David Botstein, a geneticist at the Stanford University Medical Center, sees little purpose in Dr. Venter's approach of sequencing genes blindly without knowing their function. "To say it is a substitute for the Human Genome Project is overstated by a lot," he said.

Though academic geneticists have long been skeptical of Dr. Venter's approach to varying degrees, his backers have invested $85 million in his Institute for Genomic Research, and the pharmaceutical company Smith-Kline Beecham recently paid $125 million for 10 percent of Human Genome Sciences Inc., the Rockville, Maryland, company that has the right to commercialize the institute's findings. Dr. Venter has no executive position in the company but owns shares whose current market value is about $12 million.

The U.S. government's plan is first to find chemical signposts along the vast length of the human genome, a procedure known as mapping, and then to analyze the sequence of the three billion chemical units between the signposts. This gigantic task will take until the year 2005. But while the government is still largely in the mapping phase, Dr. Venter said he has already fished out most of the genes.

His method is a shortcut around a shortcut. It focuses on just the 3 percent or so of human DNA that is used to specify proteins, the working parts of living cells. Even within this 3 percent, Dr. Venter analyzes the chemical sequence of just short stretches of genes, each about 400 base units in length.

How can such tiny flecks of DNA give any idea of the mighty canvas from which they have been scraped? By comparing them with the sequence of known genes from other species, Dr. Venter said, he can guess the natural role of about a third or more of the genes he samples. And his laboratory is churning out sequences so fast that they are beginning to overlap, so that already in a few cases the full structure of a gene can be inferred from its fragments.

In just over a year, his institute has analyzed 100,000 genetic fragments from human DNA sequences amplified in clones. These include samples of at least half of all human genes, Dr. Venter said, praising the young scientists and technicians who tend the 30 gene-sequencing machines in the institute and the attendant computers and clone-preparing robots.

"There is a tremendous esprit de corps here," he said. "People think they are taking part in history."

Dr. Venter came to genome sequencing by a circuitous route. He nearly dropped out of high school in San Francisco to devote his life to surfing. He became a champion swimmer, then a Navy medical corpsman in Vietnam. That led him to enroll as a premedical student at the University of California, San Diego, but he was deflected from medicine into biochemical research.

"I've always let the results tell me where to go next," he says. "My favorite hobby is sailing. I am always tacking."

At the National Institutes of Health, he spent 10 years isolating a protein that serves as the heart muscle cells' sensor for responding to adrenaline. Frustrated with this pace of discovery, he spent two more years sequencing long stretches of DNA but turned up only eight new genes. It was time for a different tack.

Living cells have no trouble recognizing the genetic coding regions of DNA. Transcripts of these regions, known as messenger RNAs, are prepared by the cell and then direct the synthesis of the genes' protein products. Usually these transcripts have a fleeting existence. But they can be captured and artificially copied in the test tube into their DNA counterparts, known as complementary or cDNAs. Unlike the messenger RNAs, the cDNAs are stable enough for their chemical sequence to be analyzed. The set of cDNAs from liver cells, say, represent all the genes the liver cells are expressing at the time.

When large-scale sequencing of DNA began to become technically possible, some biologists asserted that it would be more efficient to sequence just the cDNAs, but most, for the sake of knowledge, preferred to do the whole genome. Dr. Venter joined the cDNA camp. His central idea was that genes could be found and catalogued very quickly by sequencing small sample pieces of each cDNA, and that this could be done on a large enough scale to study for the first time the pattern of gene expression in different types of cells.

The task would be all the more efficient since a gene can be uniquely identified by sequencing just a short region of its cDNA. Most biologists could not see the point in identifying a gene before discovering its function. Many patent lawyers thought otherwise.

An immediate crisis was precipitated by Dr. Venter's double shortcut. In 1991, Dr. Bernadine Healy, then director of the National Institutes of Health, sought to raise Dr. Venter's budget and apply for patents on the new genes he was fast discovering lest Japanese and European companies do so first, reaping a windfall of medical and commercial applications. This brought her formidable energies into collision with another forceful scientist, Dr. James D. Watson, who was head of the NIH's half of the Human Genome Project, which is shared with the Department of Energy.

Like many other academic scientists, Dr. Watson thought wholesale patenting of unknown genes was a poor idea because it would inhibit the exchange of research information. And the Human Genome Project he headed twice turned down Dr. Venter's request for funds.

"Watson turned against me," Dr. Venter said, adding that the geneticists who run the project feared his shortcut to the genes would make Congress wonder why it was spending $3 billion to sequence the "junk" that remained. Dr. Watson said Dr. Venter's grant applications were turned down by panels of experts "at a level I couldn't reverse."

The clash of interests had unexpected outcomes. One was that in April 1992 Dr. Watson resigned from the Human Genome Project, in part because of disagreement with Dr. Healy over the patenting of DNA sequences. Another was that the dispute attracted the interest of the business community. Dr. Venter began receiving offers to leave the NIH and form a company. He accepted one from Wallace Steinberg of the Health Care Investment Corporation to form a nonprofit institute, with a budget of $70 million (now up to $85 million) over 10 years.

The conditions of this fairy-godmother deal allow him to pursue whatever scientific goals he wishes, Dr. Venter said, and to publish all his DNA sequences in scientific journals. The only restriction is that before publication, he must let Human Genome Sciences, a company formed by Mr. Steinberg, study the sequences for at least six months, and for 12 more months in the case of genes it wishes to commercialize.

The test of Dr. Venter's approach is probably yet to come. Scientists usually frown on blind gathering of data; the preferred method is to look for specific facts that will confirm or refute an idea about how nature works. Some geneticists refuse to be impressed by the sheer number of DNA sequences his laboratory has analyzed.

"No one ever doubted he could get a hundred thousand sequence runs," Dr. Botstein said. "But it doesn't mean any more now than when he first proposed it."

As for the general approach of sequencing cDNA rather than the whole genome, Dr. Botstein said, "There are other ways of going at the cDNA but none that have the same illusion of speed."

Other biologists said they believe his data will prove widely useful. "This is a fine start," said Dr. Hamilton O. Smith of Johns Hopkins University, a Nobel laureate who is a board member of Dr. Venter's institute. "It's getting a lot of valuable information much earlier than anyone imagined we would get it."

The cDNA approach is strongly defended, not surprisingly, by William Haseltine, chief executive of Human Genome Sciences. Academic critics "have a lot to learn," he said. In two or three years, he added, "the great effort to discover human genes will be over; the task then will be to redefine what a biologist does."

He said, "A biologist will start with a gene itself and match it to a phenomenon," instead of the other way around.

The success of Human Genome Sciences may be affected by the still unresolved issue of whether patents will be granted for large numbers of genes of mostly unknown function. The company's strategy, however, is not to rely on this protection but to develop full-length sequences for genes whose function has been determined. These are more likely to prove patentable. The National Institutes of Health decided recently not to pursue its patent application for the gene fragments that Dr. Venter isolated while on its payroll.

With his machines analyzing 1,000 DNA sequences a day, and a Maspar supercomputer to keep track of the streams of data, Dr. Venter seems to be surfing on top of the scientist's world. He is developing computer programs to infer what a gene does from its chemical sequence.

"All our intellectual effort is on the information side," he said. "The sequencing is totally routine."

He has several ambitious goals. He said he has already achieved one, to find a majority of the human genes, though since he has not yet published his 100,000 sequences, other scientists cannot yet assess his claim. Dr. Haseltine, for one, said he believes the sequences represent less than half the full gene set. The issue depends on how many duplications there are among Dr. Venter's sequences and on how many genes there are in all, with estimates ranging between 50,000 and 100,000.

A second goal is to describe the anatomy of gene expression in human tissues. The cells of the brain or liver or intestine presumably express different subsets of genes to perform their special functions, and these subsets should be revealed in the cDNAs analyzed from each organ.

The anatomy project will yield some important medical insights, Dr. Venter hopes, like the retinue of abnormal genes that get switched on in various types of cancer. The first appearance of these genes would serve as early detection signals. One of his colleagues is exploring the changes in gene expression that occur in aging.

But his "real goal," he said, is human evolution. "What we hope to do is to follow evolution at the genome level," he said. "We will look for evolutionary events that lead to the development of new genes." One example is the new genes for brain structure that presumably emerged with the evolution of humans from primates. That would be a leap beyond the present approach, which, because of the cost of DNA sequencing, is mostly confined to studying how a single gene varies from one organism to another.

"My motivation for leaving the government was to go ten times faster than I could have done at NIH," Dr. Venter says. "I didn't do this to become a multimillionaire. I never negotiated for stock. I got it as a matter of course because Steinberg's group offers leading scientists ten percent of the companies they form."

Dr. Venter's data on human genes may yet bring him fame, and in the pursuit he has already found fortune.

—NICHOLAS WADE, February 1994

Researchers Locate Gene
That Triggers Huntington's Illness

AFTER 10 BACKBREAKING YEARS in a research purgatory of false leads, failed experiments and long stretches of mordant despair, an international team of scientists says it has discovered the most coveted treasure in molecular biology, the gene behind Huntington's disease.

Now that they have the gene in hand, researchers say they can begin making headway in understanding the disorder, a neurodegenerative illness that usually strikes persons in their thirties or forties, insidiously destroys body and sanity alike, and kills within 10 to 20 years.

Huntington's disease afflicts about 30,000 Americans, and as many as 150,000 are at risk of developing it. The best-known victim was the folksinger Woody Guthrie.

The first clues to the gene's location came in the early 1980s, at the dawn of the contemporary era of molecular genetics. But researchers soon ran into a succession of snags that transformed the search into an irresistible if irritating quest that seduced some of the biggest names in biology.

Of particular interest to scientists, the mutation that causes the disease is one they have lately seen in genes that cause other illnesses, a sort of molecular accordion effect in which a tiny segment of the gene is abnormally expanded and repeated over and over.

Researchers emphasized today that much work needed to be done before they could use the mutation as any sort of precise prognostic tool. Nor does finding the gene mean that a treatment for the disease is imminent. But the discovery is essential to cracking the puzzle of Huntington's.

The finding, reported in the journal *Cell*, credited as its author the Huntington's Disease Collaborative Research Group. This reflects a rare instance of sustained scientific cooperation in which six laboratories in the United States, England and Wales shared their data and ideas.

Biologists everywhere, even those who were competing against the victorious consortium, greeted the news with uninhibited joy and deep relief.

"This is a landmark event," said Dr. Rick Myers of the University of California at San Francisco, who had independently been seeking to isolate the gene. "I can hardly believe that it's finally here after all these years. This is a very important, a very significant discovery."

Dr. David L. Nelson, a molecular geneticist at Baylor College of Medicine in Houston, said: "I think this is fantastic. It's taken so long to find this gene, and there's been such bizarre speculation about why people couldn't get it, that I'm relieved and thrilled to see the search has ended." Dr. Nelson has worked on another disorder, called fragile X syndrome, which is caused by a similar abnormal gene expansion.

Dr. Nancy S. Wexler of the Hereditary Disease Foundation in Santa Monica, California, and Columbia University in New York, helped hold the consortium together over the years and often traveled to towns in Venezuela where many of the villagers suffer from Huntington's. She said she had just been about to walk out the door to leave for South America when she heard the news that her group had isolated the gene.

"I felt like I'd walked into a brick wall," she said. "I was stunned. I was ecstatic. I was wandering around like a zombie after that."

Dr. Wexler, whose mother died of Huntington's disease, has a 50-50 chance of having inherited the gene. A test has been available for about a decade that predicts with high accuracy who is likely to be at risk, but Dr. Wexler declines to discuss whether she has taken the test.

The next step in research is to find out how the protein produced by the normal version of the Huntington's gene works in the body, and why the expanding mutation within the gene has such catastrophic consequences.

"From what we've seen so far, the protein doesn't look like anything else we're familiar with," said Dr. Francis S. Collins of the University of Michigan in Ann Arbor, another participant in the collaboration.

The disorder sometimes begins in childhood or adolescence, but more often is silent until well into adulthood, at which point the symptoms begin: random, uncontrollable movements in every part of the body, psychiatric disorders, mental deterioration and death. Huntington's victims can be mistaken for drunks, so careening is their walking and so slurred is their speech. Researchers now suspect that some of the Salem witches may have suffered from Huntington's.

The disease results from the extensive death of neurons in the basal ganglia, a region of the brain that controls movement and possibly cognition.

Rare as the disorder is, it has remained much in the public eye over the years, partly because the search for the gene has been so widely publicized.

Through a stroke of great luck 10 years ago, Dr. James F. Gusella of Massachusetts General Hospital and the leader of the collaboration came up with a so-called marker for Huntington's, a piece of DNA that indicated roughly where the gene must be, somewhere on the upper arm of chromosome 4, one of 23 pairs of chromosomes packed in every human cell.

Dr. Gusella and his collaborators assumed it would be a relatively straightforward task to find the specific gene, but instead scientists floundered for years as other genes—including those for cystic fibrosis, muscular dystrophy and neurofibromatosis—were plucked out to much fanfare.

The scientists repeatedly were led astray by unusual inheritance patterns of the Huntington's gene and by the complexity of working near the tip of a chromosome. The end regions of chromosomes are thought to be dense with genes and to be subject to a lot of so-called recombination, or chromosomal-exchange events, making them very difficult to sift through.

Scientists said the detection of the Huntington's gene marks the closure of the era of laboriously tracking down individual genes for diseases. From now on, biologists plan to put most of their enterprise into the Human Genome Project, the vast federal effort to systematically lay out all 100,000 genes found in the human blueprint.

At last the scientists identified a gene with all the hallmarks of being unstable and subject to dangerous expansion, exactly as they had seen in fragile X syndrome and two other hereditary disorders.

In the Huntington's gene, the mutation affects a triplet of genetic subunits, or bases, represented by the chemical initials CAG. In normal people, the gene has from 11 to 34 of these triplets. But Huntington's patients possess anywhere from 35 to 100 or more of them. This molecular stutter either disrupts the gene's ability to make a protein at all or results in a misshapen and malfunctioning protein. In either case, the defect results in the death of brain cells.

Examining 75 families with a history of Huntington's, the researchers have seen the abnormal expansion in every case of an afflicted patient. They now are firming up the evidence suggesting that the exact number of excess triplets predicts when in life a person will fall prey to the illness.

Should the correlation hold, enabling researchers to tell people that they are going to get an incurable disease and when it will strike, Dr. Collins said the knowledge could strain genetic counseling to its limits.

—NATALIE ANGIER, March 1993

Researchers Link Obesity in Humans to Flaw in a Gene

LENDING MIGHTY SUPPORT to the theory that obese people are not made but, rather, born that way, scientists have discovered a genetic mutation that is thought to be responsible for at least some types of obesity.

The mutation is believed to disrupt the body's energy metabolism and appetite control center, the mechanism that tells the brain one has eaten enough and has sufficient fat stores to meet the demands of the day. Without a working hormonal signal for fullness, or satiety, a person might continue to overeat even when extreme corpulence threatens health and ego.

The discovery could accelerate scientists' understanding of the molecular basis of obesity, a condition that afflicts one in three Americans, although a practical application is probably a decade away. People are considered obese when they are more than 20 percent above their ideal weight, at which point their risk for disorders like diabetes, high blood pressure and heart disease rises.

Eventually, the finding might lead to novel and more effective therapies for weight problems, notably a drug that would mimic the protein produced by the newly discovered gene. In theory, supplying people with the satiety signal they lack could help them feel satisfied with smaller amounts of food and thus lose weight without the sense of deprivation and hunger that often undermines conventional efforts to shed fat. Another possibility is testing for a genetic predisposition early in life, when modifying the diet would mitigate the effects of the gene.

But researchers caution that it will take at least five to ten years to translate the preliminary results into a medication. Much work remains to be done to prove that the gene, called obese, or ob for short, is indeed the body's satiety signal and that the use of it in experimental animals leads to weight loss.

Dr. Jeffrey M. Friedman of the Howard Hughes Medical Institute at Rockefeller University in New York and five colleagues reported the isolation of the ob gene in the journal *Nature*.

"I think it's a landmark paper," said Dr. Timothy J. Rink, president and head of research at Amylin Pharmaceuticals in San Diego. "It isn't all demonstrated out yet, but this looks to be the central hormonal control of body weight and body fat." Dr. Rink wrote an editorial accompanying the report.

Dr. Phillip Gorden, director of the National Institute of Diabetes and Digestive and Kidney Diseases in Bethesda, Maryland, said: "This is a major breakthrough, but it is very important that people understand it will not have any immediate application. The treatment and prevention of obesity still rest with diet and exercise."

The scientists found the gene by studying a strain of mouse that can balloon up to five times the girth of normal mice and that is also subject to diabetes and other disorders. The researchers then used the mouse gene to scan through human DNA, which led them to the human ob gene. The mouse and human genes are nearly identical, which means that the molecular mechanism for regulating fat storage and appetite has been conserved through tens of millions of years of evolution.

Whether in rodent, human or other mammal, the gene appears to be switched on solely or largely in fat tissue, where it generates a hormone-like protein that is secreted by the fat cells into the bloodstream. Once in the blood, the protein is thought to travel up to a region of the brain, called the ventromedial nucleus of the hypothalamus, considered the major controller of appetite. In that way, the fat stores of the body tell the brain how big or small they are, and the brain can regulate food intake accordingly.

"The idea is that there is a physiological pathway that controls body weight," Dr. Friedman said. "Each of us has a certain body weight that usually is stably maintained, and this requires signals to the brain about what you weigh." By this model, a mutant version of the gene would either fail to make the hormone altogether or would produce too little of it. Thus, the brain would not get the proper message on its adipose status.

Psychologists who study the lives and struggles of obese people in America greeted the new work with enthusiasm, declaring that it could help the normal-weighted be more sympathetic toward the obese.

"There's a very strong tendency to blame people for their weight and connect it with laziness, poor personal hygiene and so on," said Dr. Esther D. Rothblum, a professor of psychology at the University of Vermont in Burlington. "This research indicates that people really are born with a tendency to have a certain weight just as they are to have a particular skin color or height."

"I hope the genetic research goes a long way toward eradicating the stigma" of obesity, she added.

But those who lobby for greater tolerance toward overweight people warn that the new research could end up heightening discrimination, by presenting obesity as a disease for which science can offer a cure.

"Instead of being seen as a genetic variant, it will be seen as a genetic disease," said Sally E. Smith, executive director for the National Association to Advance Fat Acceptance, based in Sacramento, California "People might even consider aborting a fetus found to carry a predisposition toward obesity." Rather than society's becoming more accepting of weight variations in the population, she said, "the worst-case scenario is that someday genetic engineering will take over and fat people will be engineered out of existence."

Scientists said a situation like that was highly unlikely. They emphasized that many factors, both genetic and environmental, influence body weight, and ob is just part of the story. The complexity of energy metabolism makes it extremely difficult to devise any sort of easy genetically engineered "cure" for fatness.

"In the mouse, at least six different genes have been identified that contribute to obesity, and I'm confident that in humans we will find as many genes or more," said Dr. Douglas L. Coleman, senior staff scientist emeritus at the Jackson Laboratory in Bar Harbor, Maine, who predicted essential features of the ob gene 20 years ago. "Besides, in humans, there are psychological factors. Eating is a good thing to do, and some people overenjoy it."

The new work lends confirmation to theories first proposed in the 1950s that there is a feedback loop between the brain and fat tissue. Scientists first started trying to tease apart the components of the loop by grafting together the veins of different experimental animals, merging their blood supply. For example, they would take one of the genetically obese mice and surgically join it to a normal mouse, an operation that resulted in the fat mouse eating less and losing weight. Scientists postulated that the obese mouse lacked a normal blood-borne satiety signal and that through the grafting it was receiving the missing protein. Other grafting experiments suggested that the hypothalamus was the neural recipient of the hormonal signal.

In the new experiments, the scientists isolated the ob gene through elaborate cloning methods, seeking differences in genetic markers between the DNA of fat mice and of normal mice. The gene that they plucked out had several features they were seeking as appropriate for a body-weight signal: It was mutated in the obese mice, suggesting a link between the aberrant gene and their condition; it was active only in fat tissue; and it had the biochemical hallmarks of making a protein that the cells would release into the bloodstream.

Although the scientists also have the human version of the gene, they still need evidence that the gene is mutated in fat humans. One group they are considering screening for such proof are Pima Indians of south-central Arizona, who suffer from a high incidence of both obesity and diabetes. Researchers believe that the fat mutation in the ob gene might once have been of benefit to the Indians, as well as to other humans throughout history, when food was scarce and people were subject to frequent famines. Those carrying the mutant gene very likely have a more efficient metabolism and can survive longer without food. Once food becomes all too commonplace, though, the people with so-called thrifty genes are those who gain weight the fastest and suffer the consequences the soonest.

—Natalie Angier, December 1994

Discovery of Gene Offers
Clues on Deafness

A LARGE EXTENDED FAMILY in the town of Cartago, Costa Rica, has long had an unusual affliction to set against its blessings.

The affliction is a strange and incurable form of genetically caused deafness. Children born into the family have a 50 percent chance of developing the disease. They learn their fate around the age of 10 when those who have inherited a genetic mutation find that they are losing their hearing of bass notes and other low noises. By 30 they are deaf in both ears.

In a finding that may help them understand the nature of other kinds of hearing loss, too, scientists have now traced the cause of the family's affliction to a previously unknown gene. The gene helps operate the delicate hair cells in the ear that respond to sound vibrations.

In the Costa Rican family, the gene has a single change, or mutation, that was present in the family's founder, who arrived in Cartago from Spain in 1713. He suffered from this form of deafness, as have half of his descendants in the eight generations since. All the children are taught to lip-read at an early age. Many stay in Cartago because the family's hereditary deafness is well known and accepted.

"Everyone in town stands close to them and speaks clearly," said Dr. Mary-Claire King, a geneticist at the University of Washington in Seattle.

Dr. King's laboratory helped isolate the gene in collaboration with Dr. Pedro E. León of the University of Costa Rica in San José. Dr. León, a tropical biologist by training, has studied the family for 20 years and researched its genealogy. Their findings were published in the journal *Science*.

Dr. Eric D. Lynch, a member of Dr. King's laboratory, led the effort to identify the genetic mutation that caused the deafness. With only a single family to work with, and thus fewer genetic differences to go on, pinpointing the gene took six years. The gene mutation involved just one of the 3,800 chemical letters that constitute the gene's DNA.

Having worked out the sequence, or order, of chemical letters in the gene, Dr. Lynch and his colleagues searched computerized databases and found that a gene of nearly identical sequence, named diaphanous, had been detected in fruit flies.

The human version of the diaphanous gene proved to be active in many different tissues of the body. But the Costa Rican family is healthy and normal in every respect except hearing. Only in the hair cells of the ear does their slightly different form of the diaphanous gene cause any problem.

The hair cells are an extraordinarily delicate mechanism. Some cells are tuned to particular frequencies of sound, and others convert the sound into nerve signals. Because of the difficulty of this task, and the heavy mechanical and electrical stresses involved, it may be that the hair cells are less tolerant of genetic blemishes than most other tissues of the body.

"The ear seems to be peculiarly vulnerable, as evidenced by the fact that there are one hundred familial diseases of hearing and that so many people become deaf," said Dr. A. James Hudspeth, a hearing expert at Rockefeller University.

Dr. Hudspeth said that identifying the genes involved in the activity of the hair cells, as the Seattle team did, would help to explain the cell's biology and to pinpoint the things that go wrong in various kinds of hearing loss. Some 30 million Americans suffer significant hearing loss, most of which is caused by damage to the hair cells.

The genetic approach to hearing, analyzing the ear by isolating the genes that operate it, is a relatively new and growing field, said Dr. Bronya J. Keats, a geneticist at the Louisiana State University Medical Center in New Orleans. The human diaphanous gene will "add to our knowledge of how the ear works," Dr. Keats said.

The Seattle team quickly came up with a likely role for the diaphanous gene in hair cells. From its role in the fruit fly, the gene's product is known to control the assembly of actin molecules, a major structural material of living cells. Actin is particularly important in the hair cells, where it provides the structural rigidity for several parts of the apparatus that converts sound into electrical impulses.

The mutation in the Costa Rican family's version of the diaphanous gene forces an error in the way the gene is processed by the cell. As a result, the last 52 in the 1,265 units of its protein product are incorrectly formed. In most cells of the body the protein oversees the assembly of actin well enough. But evidently the hair cells are more stringent in their requirements, and the protein's misshapen tail must somehow interfere with its duties.

The researchers hope to find other genes that play important roles in the hair cells' activity, since a thorough understanding of the relevant genes should help them figure out the causes of various kinds of deafness.

—NICHOLAS WADE, November 1997

3

———————

THE NEW SCIENCE OF GENOMICS

At the beginning of the 1990s, the longest piece of DNA that had been sequenced was about 200,000 bases in length. The goal of sequencing the genomes of bacteria, the smallest forms of life capable of independent existence, seemed distant, since bacteria have genomes of a million base pairs and upward.

But a whole new field was opened up in 1995 with the development of a powerful new method for sequencing bacterial-length genomes. For the first time it was possible to describe all the genes needed by a free-living organism, a full parts list of the items required for independent existence.

In fact not all of the genes can yet be identified. The first bacterial genomes have demonstrated the extent of biologists' ignorance, since typically a third of the genes code for proteins whose role is unknown. Still, having the genomes defines the nature of the problem.

The genomes of disease bacteria are of immense interest to microbiologists, since they lay bare, at least in principle, every weakness and every weapon in the bacterium's toolkit, from its nutritional needs to how it dodges the body's immune defenses. Since bacteria around the world are becoming increasingly resistant to antibiotics, the new knowledge is opportune. Governments and foundations have taken a sudden interest in the field, and plans are now under way to sequence the genomes of most major pathogens.

Bacteria are organized differently from the cells that make up the bodies of plants and animals. The latter are known collectively as eukaryotes, meaning cells that have a nucleus, the inner compartment where the DNA is sequestered. Bacteria do not have a separate nucleus, and their DNA is anchored loosely to the cell membrane.

Archaea are single-cell organisms that resemble bacteria in some ways, eukaryotes in others. The Archaea are less well known, since they tend to live in out of the way environments, like deep-sea vents, and were recognized only recently as different from bacteria.

Eukarya, Prokarya (bacteria), and Archaea represent the three kingdoms of life on the planet, and the division between them occurred long ago, maybe close to the beginning of life. Those who study the origin of life are keenly interested in comparing the three kingdoms at the level of their genomes. This has begun to be possible now that the first archaean, *Methanococcus jannaschii,* has been sequenced, as well as the first eukaryote, the homely fungus known as yeast.

Bacterium's Full Gene Makeup Is Decoded

LIFE IS A MYSTERY, ineffable, unfathomable, the last thing on earth that might seem susceptible to exact description. Yet now, for the first time, a free-living organism has been precisely defined by the chemical identification of its complete genetic blueprint. The entire DNA sequence of a free-living organism has been deciphered, displaying a full set of the genes needed for life, two scientists announced in Washington, D.C., at a meeting of the American Society for Microbiology.

The sequence is a chain of 1,830,137 DNA bases, the chemical units of the genetic code, which constitute the entire genetic database of the bacterium known as *Hemophilus influenzae*. The microscopic organism possesses all the tools and tricks required for independent existence. For the first time, biologists can begin to see the entire parts list, as it were, of what a living cell needs to grow, survive and reproduce itself.

The result is a personal triumph for Dr. J. Craig Venter, the scientist who led the sequencing work.

Dr. Venter left the National Institutes of Health after disagreement about the best technical methods to be used in its Human Genome Project, and with private money has now won the race to sequence the first free-living organism. He first applied for government funds to sequence the *Hemophilus* bacterium, but he said he was turned down on the ground that his approach would not work.

"I think it's a great moment in science," said Dr. James D. Watson, codiscoverer of the structure of DNA and a former director of the federal project to sequence the human genome. "With a thousand genes identified, we are beginning to see what a cell is," he said.

"This is really an incredible moment in history," Dr. Frederick R. Blattner of the University of Wisconsin said in an interview. "It demonstrates the ability to take the whole sequence of an organism and work down from that to its genes, which is what geneticists have been dreaming of for a long time."

Dr. Blattner heads the National Institutes of Health project, now almost half completed, to sequence the DNA of another bacterium, *Escherichia coli.*

The importance of the *Hemophilus* sequence, in Dr. Blattner's view, is that in obtaining a full catalogue of an organism's genes, "the whole paradigm of genetics is reversed." Until now geneticists have discovered genes by seeing what function is impaired when a mutation, or change of bases, is made in a bacterium's DNA. With the full catalogue of genes in hand, they can start with a gene and search for its function.

Full genome sequences will also open the door to medical applications, like pinpointing a bacterium's virulent genes by comparing its harmless and disease-causing forms.

Hemophilus—no relation to the flu virus—colonizes human tissues, wherein its virulent form can cause earaches and meningitis. The task of sequencing *Hemophilus* was suggested by Dr. Hamilton O. Smith of Johns Hopkins University School of Medicine, who won a Nobel Prize for discovering special enzymes it possessed. Dr. Smith proposed a strategy for sequencing the bacterium and prepared a library of clones, or chopped up and amplified pieces of DNA, which he gave to Dr. Venter's laboratory, the Institute for Genomic Research now in Rockville, Maryland.

Although many viruses have been sequenced, their gene sets, or genomes, are quite small, since they replicate themselves by usurping the machinery of living cells and lack the genes for independent existence. Smallpox virus, for instance, has a genome of only 186,000 DNA bases. The *Hemophilus* genome sequenced by Dr. Smith and Dr. Venter is nearly 10 times as large. Their article about their work appeared in the journal *Science.*

Dr. Venter said the sequencing took him just under a year. Genomes this big are usually tackled by first making a "map" of known chemical signposts along the DNA chain, and then sequencing between the signposts. He said his progress was so rapid because he had skipped the time-consuming mapping stage and relied on software programs he had developed to fit together the numerous pieces of the enormous jigsaw puzzle he had created.

With the full DNA sequence in hand, Dr. Venter has started to analyze it, although "it will take all of us months, if not years, to truly understand it," he said.

The *Hemophilus* genome seems to contain 1,749 genes. By comparing the sequence of these genes with those of known function from other organisms, Dr. Venter has predicted the biological role of most of them. They fall into 14 major categories that must include "all the enzymes necessary for life," Dr. Venter noted. Many of these genes are clustered into logical groups, corresponding to the series of enzymes needed at each step of a biochemical pathway.

Another feature of interest to microbiologists is the presence of a latent phage, a virus that infects bacteria, whose DNA sequence is nestled inside that of *Hemophilus*. The bacterium also has six identical sets, spread around its chromosome, of the genes that specify ribosomes, part of the cell's protein-making machinery, Dr. Venter said.

Deciphering the first full gene set of an organism like *Hemophilus* is a landmark in the young science of genomics, the study of living things in terms of their full DNA sequences or genomes. Even the simplest organisms possess a daunting length of DNA. Only recently have machines been developed that automate the task of identifying the sequence of As, Ts, Gs and Cs, short for adenine, thymine, guanine and cytosine, the chemical groups that make up the four-letter alphabet of the genetic code. The machines can handle fragments of DNA that are a few thousand letters in length, but it is no simple matter to reassemble the fragments in the correct order as they exist in the organism's genome.

Each human cell contains a copy of the human genome, which has about three billion base pairs and would make an invisible skein six feet in length if all the DNA were to be unpacked from its tightly wound coils in the cell's nucleus. A genome that size lies way beyond the present reach of sequencing machines and software programs, although some scientists now think it can be completed within a decade.

The longest genomes sequenced before this were those of various viruses, which range up to about 200,000 base pairs in length. To reproduce, viruses simply hijack the biological machinery of the animal or bacterial cells they parasitize; hence, their miniature genomes, while of great interest to virologists, reveal little about the characteristics of independent, free-living cells.

Deciphering the *Hemophilus* bacterium's genome is a milestone in sequencing technology because its DNA is almost 10 times as long as that of the typical virus. Moreover, it was accomplished by a strategy so adventurous that an expert advisory committee of the National Institutes of Health, to which Dr. Venter and Dr. Smith had applied for money, turned down their approach as unworkable.

Dr. Venter's achievement threatens to make him part of the scientific establishment with which he has long been at odds because of his liking for shortcut approaches to genome sequencing that other experts say are unlikely to work.

In the case of *Hemophilus,* Dr. Venter said he applied for money to the National Center for Human Genome Research, a part of the National Institutes of Health. At the same time he started work on the project with other money he had available. When he had completed 90 percent of the sequence, he received a letter saying his application had been turned down, since the committee of experts that reviewed it felt he would be unable to close the remaining gaps between his assemblies of shorter sequences.

This well-known problem in sequencing arises because there are usually a few genes that produce proteins that are toxic to the cells in which they are growing, causing the puzzle to have missing pieces. But Dr. Venter sidestepped this problem with a second set of clones that did not produce proteins.

Dr. Robert L. Strausberg, an official who oversees the committee that recommended against Dr. Venter's application, confirmed that it was turned down because of concerns like that of gap closure. But he denied that the committee had made an error of judgment.

The strain of *Hemophilus* sequenced by Dr. Venter and Dr. Smith turned out to contain 1,830,137 base pairs of DNA, arranged in a single circle. The sequence of the base pairs constitutes information that codes for 1,743 genes.

Since life is a unity, at least at the level of DNA, a gene of a certain DNA structure in one organism is likely to play the same biological role as a gene of similar structure in another organism. The Venter-Smith team has been able to assign probable functions to 1,007 of *Hemophilus*'s genes by comparing their DNA sequences with those of already sequenced genes of known function from various other organisms.

"You can see how the organism replicates itself, translates proteins, deals with the environment," Dr. Smith said. "All the machinery is right there in front you. You can see that it uses six different sugars because the genes are right there, and that it does not metabolize certain other sugars."

Of even greater interest are the 736 genes for which no role could be assigned after data bank searches. These presumably have novel functions that, once discovered, will take biologists substantially closer to a complete understanding of *Hemophilus* and similar bacteria.

Biologists' data banks now contain the DNA sequences of many genes from many different organisms. That the role of so many of *Hemophilus's* genes could be guessed through their sequence similarities suggests either that a limited number of gene-protein motifs have arisen in the course of evolution, or at least that once nature has developed a theme that works, it exploits it in myriad ways.

The study of evolution at the level of DNA is perhaps the greatest prize that the *Hemophilus* genome promises. By itself, it is just one of countless bacterial species. But when its genome structure can be compared with those of other organisms whose DNA sequences are now nearing completion, biologists will gain a deep insight into the differences between the major branches of life and how those differences evolved.

Just as astronomers with powerful telescopes can see very distant events that happened close to the birth of the universe, so, too, biologists who analyze genomes can hope to infer events that date from the earliest moments in the evolution of life.

One of those events was the division of the first living cells into three major kingdoms, known as Archaea, Prokarya (bacteria), and Eukarya. The Archaea is an ancient group of bacteria-like organisms found in the hottest places on Earth, a possible hint as to where life may have originated. The Eukarya includes multicellular organisms like plants and animals.

The *Hemophilus* genome, representative of the bacteria, which are also known as prokaryotes, triangulates one point of this fundamental division. Dr. Venter expects to have soon completed the genome sequence of *Methanococcus jannaschii*, a member of the Archaea. Other biologists working on yeast, a eukaryote, expect to complete its sequence of 12 million bases within the next several months.

Dr. Walter Gilbert, a Harvard biologist, said that completion of these genomes will "give us a very deep way of looking back at the original division between the Prokarya and Eukarya and between them and the Archaebacteria."

Comparisons aside, the *Hemophilus* genome has considerable intrinsic interest. The bacterium lives in humans, its sole host, and one of its major problems is that its host's immune system is bent on killing it with antibodies tailored to attack its various components. *Hemophilus* has two cunning genetic survival strategies, the full extent of which has become evident with the sequencing of its full DNA.

One is a genetic signature tune that enables it to recognize and preferentially absorb fragments of its own DNA from its surroundings, ignoring those shed by human cells or other bacteria. The new fragments can be incorporated into the genome, giving some bacteria a genetic advantage.

The signature is a nine-base sequence of DNA—AAGTGCGGT—that Dr. Smith and his colleagues have found is repeated nearly 1,500 times within the *Hemophilus* genome. The bacterium presumably has receptor proteins on its surface that recognize and pull in any pieces of DNA from dead and dying cousins that have this unique sequence. The scavenged pieces of DNA "are genetically recombined into the recipient, and that allows it to play combinatorial games with all the mutations within the population, so it increases the possible diversity," Dr. Smith said.

Hemophilus's other survival strategy is a special stretch of DNA, known as a tetramer repeat unit, that fosters errors in the bacterium's genetic copying process. Though that might not sound like an advantage, the error means that any gene with a tetramer repeat may be active in some bacteria but not in others. The purpose may be that within a population of *Hemophilus* bacteria, some members can conserve energy by not expressing most of these special genes, but a few members will have the genes in active form, and they and their progeny will rapidly become predominant in circumstances in which the gene confers some advantage.

Three such *Hemophilus* genes with tetramer repeat units had already been identified by E. Richard Moxon, a molecular biologist at Oxford University. With the full genome in hand, Dr. Moxon has found eight more such genes, Dr. Smith said. Some are adhesion genes, meaning that they specify a protein that helps the bacteria to bind to the surface of the lung or throat and avoid getting washed away. Three of the genes produce proteins that help *Hemophilus* hoist in vitally needed atoms of iron, an element that the body tries to keep inaccessible as an antibacterial defense.

Dr. Smith believes that the study of bacteria, once the first wave of molecular biology but now something of a backwater, will be revitalized as more bacterial genomes become available and new insights are gained into how each species of bacteria exploits its ecological niche. In particular, the new understanding of bacteria will allow vaccine makers to design genetically crippled bacteria to serve as a superior class of live vaccines. These may be needed to replace antibiotics, whose power is now fading as bacteria are steadily evolving resistance to them.

—NICHOLAS WADE, May 1995

Thinking Small Paying Off Big in Gene Quest

WORKING IN THE SHADOW of the vast project to decode human genes, biologists are rapidly deciphering the genetic makeup of much smaller organisms, including microbes that cause disease.

The result, these scientists say, is that the full DNA, or genomes, of many pathogens is likely to be decoded in the next several years, offering new drug and vaccine strategies.

They say their field will have an impact far sooner than will the better known $3 billion project to decipher the human genome, which began in the late 1980s and is not expected to be completed before 2000.

"It is absolutely clear that the availability of whole genomes has changed the way we are doing science," said Dr. E. Richard Moxon, an infectious disease specialist at Oxford University.

Aside from their medical importance, the small genomes are also of interest in two other arenas. Biologists hope that by comparing diverse genomes, they will be able to trace the tree of evolution back to the origin of life, or at least to its earliest branches. And industrial chemists are screening genomes from the poles to the deep sea in search of enzymes with special properties.

Sequencing a genome means determining the nature and order of all the chemical building blocks, each represented by a letter, of an organism's DNA. The smallest known bacterial genomes contain as few as 500,000 letters in their genetic instruction set. That is just a few pages, compared with the three-billion-letter tome of human DNA, but it is still daunting enough that the first bacterial genome was sequenced only two years ago.

Sequencing a small genome costs $1 million to $15 million, depending on the length of the organism's DNA. Despite the costs, which are enormous by the usual standard of biological research, plans are now under way for sequencing the organisms that cause diseases like malaria, syphilis, Lyme disease, typhus, gastric ulcers and gonorrhea.

The agencies financing the work are so eager to stake out claims that the National Institutes of Health and the Wellcome Trust of London each formed plans to sequence the tuberculosis bacterium before deciding to join forces instead.

To the human eye, the printed result of a computer-generated genome sequence is a meaningless string of As, Gs, Ts and Cs, representing the four chemical components of the genetic code. Sequencing has long been regarded by biologists as about as thrilling as running an index fund, but the power of the computer has begun to modify that.

Computers can annotate the text of a genome, marking where its genes probably start, flagging the on and off gene switches that are found in DNA regions between the genes and marking the hidden viruses that have slipped their DNA into their host's genome. Most important, computer programs can identify the likely role of many genes by comparing their DNA sequence with the thousands of genes of known function whose sequences have now been deposited in data banks.

Genomes annotated in this way are transformed from gibberish to draft blueprints of a living organism. For scientists, the annotated genomes of pathogenic bacteria are like a decrypted top-security message about the enemy's strategy and tactics. They lay bare every inherited weapon and defense, even though much is not yet understood.

The era of small-genome sequencing began in 1995 when *Hemophilus influenzae,* a bacterium that causes ear and throat infections (and is no relation of the flu virus), was sequenced by a team led by Dr. J. Craig Venter, of the Institute for Genomic Research, now in Rockville, Maryland, and Dr. Hamilton O. Smith, of Johns Hopkins University School of Medicine in Baltimore.

The availability of the *Hemophilus* genome offered several deep insights into the microbe's game plan, including the crucial mechanism by which it keeps shifting the composition of its coat to evade its host's immune system. News of the sequence had particular impact because few had believed a bacterium could be sequenced so easily; on that advice, the National Institutes of Health had refused to finance the project. Showing that bacterial genomes now lay within the sequencer's reach set off something of a researchers' gold rush.

At a conference on small genomes held at Hilton Head, South Carolina, Dr. John Stephenson, a senior official of the Wellcome Trust, said: "It is incredible to realize that two years ago, everyone was very skeptical that Craig Venter could do what he did. The world of microbial genomics was changed overnight."

The trust, which has built a $13 billion endowment through the recent sale of its shares in the Burroughs Wellcome Company, decided within a few months of the publication of the *Hemophilus* genome to set aside $25 million for work on pathogenic genomes, Dr. Stephenson said. Dr. Anne Ginsberg, of the National Institutes of Allergy and Infectious Diseases, said the *Hemophilus* genome success had caused a "change in the landscape." Her agency is paying for the sequencing of several pathogenic genomes.

Dr. Venter noted that two genomes had been completed in 1995 and four last year and that as many as eight of the thirty-five genomes now being sequenced might be finished this year. He said that with sufficient financing, the genomes of all the major human pathogens could be sequenced by the end of the decade and that small genomes would be the major source of the half million new genes he predicts will be discovered by then. The human genome has far fewer genes per length of DNA than the very compact genomes of bacteria, Dr. Venter and others said, but it is likely to prove to be far harder to interpret.

In a telephone interview, Dr. Harold E. Varmus, director of the National Institutes of Health, commented that both kinds of genome work would have significant impact. "I don't think it is fair to say one will be bigger than the other," he said.

The speakers at the conference reported progress with a wide range of small genomes, mostly with enthusiasm, although one exhausted participant declared that he would never sequence a genome again.

Dr. Frederick R. Blattner of the University of Wisconsin announced a long-awaited triumph: the completion of the sequence of *Escherichia coli*, the standard laboratory bacterium of molecular biology. While *Hemophilus* has just over 1.8 million units, or pairs of chemical bases, the *E. coli* genome is 4,638,858 base pairs, which contains the instructions for 4,286 genes.

The probable roles of 1,817 of these genes have been identified by comparing them with genes of known function in databases. Even though biologists have studied the minutest parts of *E. coli* for decades, the roles of its other 2,469 genes are unknown. "That shows the limitations of bottom-up genetic approaches," Dr. Blattner said, "even when applied over extended periods."

For much of his five-year struggle—the National Institutes of Health nearly lost confidence in the project at one point—Dr. Blattner and his team were competing with a Japanese team now led by Dr. Hirotada Mori, of the Nara Institute of Science and Technology. Dr. Mori said he had completed his *E. coli* sequence a week after Dr. Blattner. The teams sequenced slightly different strains, so their sequences would not be expected to match exactly. "As a competition, it was perfectly dignified," Dr. Blattner said, "but I'm glad that, like the Packers, we won."

As work on one genome after another was described at the meeting, the scientists' mood was like that of people looking at newly discovered treasure maps, with the treasure not yet in hand but with wonderfully tantalizing clues all about.

For example, the order of genes in a genome seems to vary widely, even between closely related species of microbes, as if evolution were constantly shuffling the deck.

A major goal of those studying pathogenic organisms is to understand the microbes' ability to evade attack by the immune system. Dr. Moxon has made considerable progress with *Hemophilus,* finding that it weaves its coat from a material made of fat and different sugars. The core of the sugar assemblies stays the same, but different kinds and lengths of sugars can be tacked onto it, depending on which subset of the 25 coat-making genes is active.

Because the sugar coats of an invading *Hemophilus* army keep shifting, the body's immune system never has a fixed target. The unchanging sugar core seems like a good target for a vaccine. Although Dr. Moxon started this work with conventional methods, he said, the availability of the genome sequence was a "wonderful gift" that allowed him to proceed much faster.

In April 1996, at a cost of $40 million, a consortium of laboratories, most of them European, completed the genetic sequence of yeast, a 12 million base-pair genome of high interest because it belongs to the realm of plants and animals and was the first such genome to be sequenced. Despite the many years yeast has been studied, more than 30 percent of its 5,885 genes turned out to be of unknown function. Laboratories in both the United States and Europe have begun major programs to identify the role of those genes.

Once there were only two ways of doing biology: in vivo, with a live organism, and in vitro, inside laboratory glassware. Biologists who now use computers to romp through DNA databases and analyze genes believe that they have struck on a third kind: biology in silico, referring to the computer's silicon chips.

New techniques sometimes turn out to be a little less sparkling than the dreams in their proponents' eyes, and traditional microbiologists expect to see a gap or two opening between the in silico discoveries and the in vivo reality. Still, attacking pathogens through study of their genomes is likely to offer a new and fundamental way to design drugs and vaccines. Dr. Waclaw Szybalski, of the University of Wisconsin, said the current methods, based a lot on trial and error, "are like trying to repair a TV set without a blueprint."

—NICHOLAS WADE, February 1997

Scientists Map a
Bacterium's Genetic Code

IN A TOUR DE FORCE of computer-aided biology, scientists have decoded the full genetic instructions of the bacterium that causes ulcers and other stomach disease and have figured out many of its cunning strategies.

The advance is likely to lend new impetus to research on the bacterium, *Helicobacter pylori,* which is a leading cause of human illness. The microbe is thought to live in almost half the world's people, though usually without causing disease. In the United States it is found in 30 percent of adults and more than half of the people over age 65, with a prevalence in lower socio-economic groups.

The *Helicobacter* genome has been deciphered by Jean-François Tomb and a team of scientists at the Institute for Genomic Research in Rockville, Maryland, directed by J. Craig Venter.

Possession of the microbe's full instruction set, or genome, gives researchers invaluable information. Like a general who learns the enemy's order of battle, researchers now stand to know everything the organism can do and how it does it. The knowledge will help them to understand which strains cause disease and to design new drugs and vaccines.

"I think it will have tremendous significance and will enable studies in many important directions," said Dr. Martin Blaser, director of infectious diseases at Vanderbilt University.

The bacterium was suggested as the agent of stomach ulcers as recently as 1983 and is now thought to cause 90 percent of such cases. Conventional medical wisdom until then and for a decade afterward held that excess stomach acid, induced by stress, was the cause of ulcers, and that theory, indicating treatment based on reducing stomach acid, created a market for two of the world's best-selling drugs, Tagamet and Zantac.

The surprising upset of established theory was accomplished after work by two Australian physicians, Barry Marshall and Robert Warren, and led directly to therapy based on antibiotics. But the antibiotics are expensive, especially in developing countries where infections often recur, raising the need for different and better treatments.

The ability to decode a bacterium's genome is a recent technical achievement that has not yet become routine. The *Helicobacter* genome is the fifth bacterial genome to be published and those of a dozen others, mostly pathogens, are in various stages of completion. Biologists expect that when a critical mass of deciphered genomes is available, many details about the armament and evolution of bacteria will become apparent.

Dr. Venter said he started the project two years ago because a company, Genome Therapeutics, of Waltham, Massachusetts, claimed it had completed the DNA sequence but declined to publish it. The claim turned out to be incorrect, he said, but also made it hard for his nonprofit institute to raise money for its own effort.

Dr. Bernd Seizinger, chief scientific officer of Genome Therapeutics, said he welcomed Dr. Venter's sequence, which is of a different strain of the bacterium. Dr. Seizinger said the company had completed 99 percent of the sequence, all that was necessary for commercial reasons, as part of a $22 million contract with Astra AB of Sweden, but could not publish the sequence because of its proprietary value.

Dr. Venter's institute bore the cost, $1 million to $2 million, out of its own money. The genome, published in *Nature* and posted on the Internet (*http://www.tigr.org/tdb/mdb/hpdb/hpdb.html*), is being made freely available.

"We paid for it ourselves and are giving it away for free," Dr. Venter said. "We are hoping it will attract some funding for us in the long run."

The bacterium turns out to have 1,667,867 units of DNA, the chemical that embodies the genetic code, arranged in a single, circular chromosome. Arrayed along the ring of DNA are the coding sequences for 1,590 genes.

The team lead by Dr. Tomb has figured out the role of many of these genes by delving into computer databases that record the already sequenced DNA of genes of known function from other organisms. By comparing the *Helicobacter* genes with those already on file, he has guessed what many are designed to do and the overall strategies that they enable the microbe to perform.

This computer-based approach, working from gene to function, is the opposite of the microbiologist's usual tactic, which is to study some property of a microbe and work backward to the gene that underlies it. The computer approach is also immensely faster because it lays bare the organism's bag of tricks at a single stroke.

Helicobacter is a remarkable microbe because it thrives in extraordinarily inhospitable surroundings, the acid environment of the stomach. To avoid being washed away, it needs to burrow into the stomach wall and stick itself to cells. There it must fight off relentless attack by the immune system.

Dr. Tomb's team has found sets of genes that perform all these duties. A batch of transporter genes makes proteins that nestle in the bacterium's cell wall and pump out acid. Another set of genes pumps in iron, a vital ingredient for bacteria and one that is particularly scarce in the stomach.

There are genes that build the powerful tail with which *Helicobacter* propels itself, a large family of genes that make glue-like proteins for sticking on to human cells, and genes that mimic certain human proteins. *Helicobacter* also possesses a cunning genetic mechanism, called slipped-strand mispairing, that continually changes the texture of its coat to keep one step ahead of the chemicals designed against it by the human immune system.

The bacterium behaves differently in different people. Most people harbor the bacterium for years or decades, and about 20 percent go on to develop gastric diseases like ulcers, said Dr. Douglas E. Berg, who studies the molecular genetics of *Helicobacter* at Washington University in St. Louis.

"There is a huge diversity of *H. pylori* strains, which may account for why some people are infected and others not," Dr. Berg said. "Having one complete genome sequence gives you a framework to look for differences."

H. pylori has probably infected humans and their predecessors for millions of years. The diseases it causes, gastritis, ulcers and possibly some stomach cancer, may arise from some change in modern human living habits that has interfered with human populations' age-old adaptation to the bacterium, Dr. Blaser of Vanderbilt suggested.

"I think we are just at the beginning of understanding *H. pylori* and its relationship to human disease," he said.

Helicobacter, Dr. Blaser noted, "is one of those interesting organisms that is on the border" between being a pathogen and a benign occupant of the gut. One hypothesis, he said, is that because of modern hygiene, people may acquire the organism later in life than used to be the case, and therefore be less tolerant of it and more prone to develop disease. "Having the genome will allow us to test this and many other hypotheses," he said.

—NICHOLAS WADE, August 1997

DNA of Organism in Lyme Disease Is Decoded

GAINING A MAJOR NEW INSIGHT into the nature of Lyme disease, biologists have decoded the entire DNA of the bacterium that causes the infection.

Somewhat like obtaining the enemy's order of battle, the achievement yields the genetic instructions for every survival stratagem the organism has developed in the course of evolution. Biologists have a new starting point from which to develop diagnostic tests and vaccines, even though they are far from understanding all the information they have obtained.

"It is certainly very useful to have the whole book in your hand, even though you can't yet read it," said Dr. Alan G. Barbour of the University of California at Irvine, who first discovered how to grow the picky bacterium in the laboratory.

Dr. Benjamin J. Luft, a Lyme disease expert at the State University of New York at Stony Brook, said the new DNA sequence would be "invaluable for the development of new therapeutics and for understanding the development of the disease."

The DNA of the organism, known as *Borrelia burgdorferi,* was sequenced by Dr. Claire M. Fraser and a team of biologists at the Institute for Genomic Research in Rockville, Maryland, and other institutions. Their report was published in the journal *Nature*.

Lyme disease, which is particularly prevalent in New York State and New England, is spread by a tick that feeds on such animals as mice, deer, dogs and humans at various stages of its life cycle. Though the disease is easily treated if antibiotics are given promptly, it can cause serious symptoms if unrecognized. A vaccine is now undergoing clinical trials, but no accurate diagnostic test is available.

Knowledge of the bacterium's full DNA sequence should help toward developing second generation vaccines, if needed, and accurate diagnostic tests. From its genetic repertoire biologists should also be able to gain a better understanding of *Borrelia's* complicated life cycle.

"It's very exciting for us, given the amount of money we have put into the field and the spotty progress in terms of the research," said James H. Handelman, executive director of the Mathers Charitable Trust of Mount Kisco, New York, which financed the research. The trust, which supports basic biomedical research, became interested in Lyme disease 10 years ago, in part because many of its board members had been exposed to it.

The genome of the Lyme disease organism turns out to consist of some 1,444,000 base pairs of DNA, the hereditary material. Most bacteria have genomes in the form of a single circular chromosome. *Borrelia's* is very unusual because it has a single linear chromosome and numerous small strips of DNA known as plasmids. Exchange of plasmids is the usual way by which bacteria transfer antibiotic resistance genes among one another.

Though few of *Borrelia's* genes have been directly studied, biologists can assign likely roles to many through their similarity to genes of known function from other organisms. Of the 853 genes on *Borrelia's* main chromosome, the roles of 59 percent have been identified. But only 16 percent of the 430 plasmid genes are familiar, Dr. Fraser's team reported.

Borrelia seems to have most of its basic housekeeping genes lined up on its main chromosome, while the plasmids carry numerous genes for making lipoproteins, substances that form the bacterium's coat. The parasite has to survive attacks by the immune systems of all its different hosts, and may do so by rapidly switching between its repertoire of lipoprotein genes so as to change the composition of its coat.

Borrelia also has many genes thought to be involved in responding to chemical cues in its environment. This is not surprising, given that the bacterium must move at the right time from the tick's stomach to its mouth parts, and that once inside its new host it must seek out various organs to lie low and avoid the immune system's attack.

Dr. Fraser said she expected that the bacterium was able to switch on one set of genes for living in its insect host and another set for living in mammals. But to her disappointment, she has not yet managed to identify any such master gene.

Dr. Fraser's team was not alone in sequencing the *Borrelia* genome. Biologists at the Brookhaven National Laboratory started to decode the organism's DNA in January 1996, in part to test a novel sequencing method. Dr. Fraser's team started in March 1996, using a method developed at the Institute for Genomic Research.

"It's like going jogging—you both jog faster if someone else is with you," Dr. John J. Dunn of the Brookhaven group said.

But Brookhaven's new method turned out not to be competitive, and Dr. Fraser's team finished first by a considerable margin.

The Lyme vaccine now being tested is based on a component of *Borrelia*'s coat known as OspA, for outside protein A. The protein was chosen because it was the most common in laboratory cultures of *Borrelia*.

Biologists now know that OspA is produced only when the bacterium is living inside its insect host. As soon as *Borrelia* senses that it is being transferred to a mammal bitten by the tick, it switches to making another protein, OspC. By sheer luck, the OspA-based vaccine seems to be effective because it starts attacking the bacteria the instant they emerge from the tick, and before they have been able to switch coats.

Borrelia is the eleventh bacterium to have its genome sequenced, and the sixth to be sequenced by the Institute for Genomic Research, which pioneered in this field in 1995.

—NICHOLAS WADE, December 1997

4

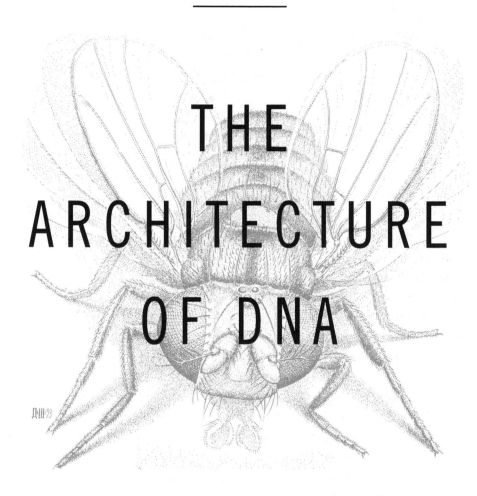

THE ARCHITECTURE OF DNA

D NA carries all the instructions needed to design, operate, maintain and replicate an organism. But this thread of life is far more than just a biological computer tape. It is twisted, coiled and packaged in ways that are essential to its functions, but that also lie at the edge of present understanding.

As noted in the first article, the DNA—six feet of it in the nucleus of each human cell—must be tightly packed to keep it in place and under control. Yet it must also be widely accessible to the cell whenever a particular gene needs to be copied. How the cell manages this feat of nonstop packing and unpacking is not known in any detail.

But it is known that some of the proteins that manage access to the DNA can physically bend it, opening the way for the machinery that copies special segments of DNA. These segments are processed into the actual instruction tapes that drive the cell's protein-making machines, ordering proteins of a specific type.

Another difference between DNA and a computer tape is that DNA is continually changing. The mechanisms of change include mobile segments of DNA that have acquired the ability to copy themselves and insert the new copies at other sites in the genome. This process has gone on for so long that the vast majority of human DNA is now made of these repetitive sequences. It is not yet known whether the repetitive DNA confers some positive benefit to the genome or is neutral in its effects.

Though DNA is made accessible when the cell needs to copy certain genes, most genes in mature cells are permanently inaccessible. The reason is that as an embryo develops, its cells become increasingly committed to specific roles, such as being a rod cell in the retina, or a hormone-producing cell in the pituitary gland. The choice of a cell's fate is sealed by the permanent switching off of all genes not needed by that type of cell.

The process of becoming committed to a specific fate is known as differentiation. Biologists would love to understand how the master plan of differentiation unfolds and how the various regions of DNA are marked so as to designate them open or closed to the differentiated cell.

Recently the subject of cloning has been in the news because of the reported generation of a sheep from the adult cell of another. The same result has since been obtained more convincingly in experiments with mice.

Cloning is routine in bacteria because they divide asexually, with parent and daughter cells having identical DNA. The trick is harder in animals because the mysterious process of differentiation must be reversed in the cell that is to be cloned.

Biologists use a method called nuclear transfer, in which the nucleus from an ordinary cell is substituted for the oocyte's own nucleus. However, until the advent of Dolly the sheep, this method only worked when the nucleus came from cells of a fetal animal. Fetal cells have apparently journeyed so short a way down the path of differentiation that their progress can be reversed when the nuclei find themselves back in the environment of the oocyte. A viable whole animal can in this way be grown from the fetal cell of another. Since the two animals are genetically identical (apart, of course, from their mitochondrial DNA), they are like twins that happen to be of different ages.

Animal breeders would in some cases like to clone from adult, differentiated cells, since that would be one way of making many copies of a champion pig or sheep. Were human cloning to be found ethically acceptable, it is easy to imagine infertile couples contemplating having children that were cloned from their own bodies.

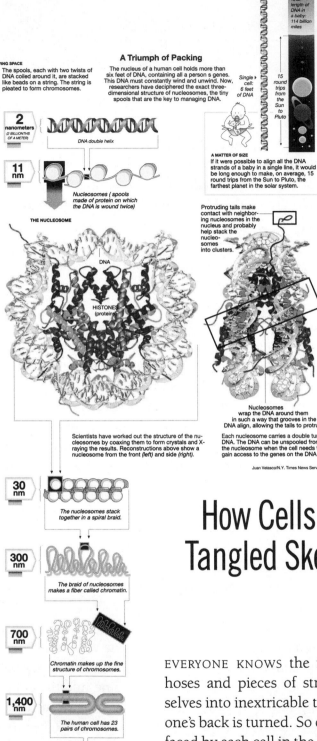

SAVING SPACE

The spools, each with two twists of DNA coiled around it, are stacked like beads on a string. The string is pleated to form chromosomes.

A Triumph of Packing

The nucleus of a human cell holds more than six feet of DNA, containing all a person s genes. This DNA must constantly wind and unwind. Now, researchers have deciphered the exact three-dimensional structure of nucleosomes, the tiny spools that are the key to managing DNA.

Total length of DNA in a baby: 114 billion miles

Single cell: 6 feet of DNA

15 round trips from the Sun to Pluto

2 nanometers (2 BILLIONTHS OF A METER)

DNA double helix

11 nm

Nucleosomes (spools made of protein on which the DNA is wound twice)

A MATTER OF SIZE
If it were possible to align all the DNA strands of a baby in a single line, it would be long enough to make, on average, 15 round trips from the Sun to Pluto, the farthest planet in the solar system.

THE NUCLEOSOME

DNA

HISTONES (proteins)

Protruding tails make contact with neighboring nucleosomes in the nucleus and probably help stack the nucleosomes into clusters.

Nucleosomes wrap the DNA around them in such a way that grooves in the DNA align, allowing the tails to protrude.

Each nucleosome carries a double turn of DNA. The DNA can be unspooled from the nucleosome when the cell needs to gain access to the genes on the DNA.

Scientists have worked out the structure of the nucleosomes by coaxing them to form crystals and X-raying the results. Reconstructions above show a nucleosome from the front *(left)* and side *(right)*.

Juan Velasco/N.Y. Times News Service

30 nm

The nucleosomes stack together in a spiral braid.

300 nm

The braid of nucleosomes makes a fiber called chromatin.

700 nm

Chromatin makes up the fine structure of chromosomes.

1,400 nm

The human cell has 23 pairs of chromosomes.

15,000 nm

Sources: Nature; Molecular Biology of the Cell, Alberts, et al. (Garland)

Chromosomes are packed into the cell nucleus.

How Cells Unwind Tangled Skein of Life

EVERYONE KNOWS the innate propensity of hoses and pieces of string to weave themselves into inextricable tangles just as soon as one's back is turned. So consider the problem faced by each cell in the human body: It must manage more than two yards' worth of computing tape in the form of delicate ribbons of DNA, the genetic material.

Although each cell is one fifth the size of what the eye can see, it has to keep its six feet of DNA neatly coiled and free of knots, tightly packed into the small compartment that is the cell's nucleus.

And packaging is only the start of the cell's problem. Its DNA is a dynamic data bank; the tape is continually being unwound and rewound at thousands of different sites as genes are made accessible and their information is copied by the cell's transcription machinery. And when the cell needs to divide, the entire tape must be split apart, duplicated and repackaged for each daughter cell.

No one knows exactly how cells solve this topological nightmare. But the solution clearly starts with the special spools on which the DNA is wound. Each spool carries two turns of DNA, and the spools themselves are stacked together in groups of six or eight. The human cell uses about 25 million of them to keep its DNA under control.

In an achievement that will lead to further understanding of how DNA is managed by cells, the fine structure of the spools and their DNA coils has now been worked out by biologists at the Swiss Federal Institute of Technology in Zurich. They mapped the three-dimensional position of individual atoms in the spool-DNA complex by the technique of X-raying the molecules in crystalline form.

Earlier work by these and other scientists had shown the general shape of the spools. The new structure shows the spools at higher resolution and delineates important new features of their organization.

"My feeling is that this is a really terrific achievement," said Dr. Wayne A. Hendrickson, an expert on molecular structure at Columbia University. Management of DNA is "one of the cornerstones of biology," he said, and "this high resolution picture has so many wonderful new ideas embedded in it."

Dr. Daniela Rhodes of the Laboratory of Molecular Biology in Cambridge, England, said the new structure "explains how the DNA is held by the protein core," adding, "The architecture of the protein is quite beautiful and very ingenious, and it gives insight into how the DNA might unwrap from the core."

The leader of the Swiss team, Dr. Timothy J. Richmond, has devoted much of his career to studying the spools, called nucleosomes, and in particular to preparing crystals of sufficient regularity to yield suitable X-ray pictures. "I have been working on this problem nonstop for eighteen years," Dr. Richmond said in a telephone conversation from Zurich. "It has affected my life for eighteen years. I have been driven to do it."

Trained in the United States, Dr. Richmond began work on the nucleosome structure at the Cambridge laboratory, and by 1984 had obtained low-resolution X-ray pictures of nucleosome crystals. Realizing that he needed better equipment to see the nucleosome at higher resolution, he moved to Switzerland, where he had access to a powerful X-ray source.

It has taken the last 13 years to prepare crystals of the right kind to yield suitable X-ray pictures. During that time, Dr. Richmond succeeded in keeping his laboratory financed and ahead of other scientists working on the problem. "We were always worried about competitors," Dr. Richmond said in an interview. "But the main driving force is the desire to see what the thing looks like. Fortunately, I was able to infect my colleagues, and they became as excited about it as me. As I talk to you it begins to sink in that we actually did it."

One of his colleagues is his wife, Dr. Robin K. Richmond, who has worked with him for eight years. She said the project had been "quite a struggle," marked by "several times when it looked like the answer was nearly there."

The nucleosome spools are made from proteins called histones, which are notable because in plants and animals their structure has remained almost unchanged throughout evolution. Eight histone molecules join together to form a spool, and the DNA is bent twice around it in a manner that stretches and overtwists the DNA helix.

The new pictures explain for the first time the purpose of the overtwisting. Viewed from outside, the double helix has a major and a minor groove because its two DNA strands are not evenly spaced but bunched together as they spiral round the helix. The overtwisting aligns the minor grooves on neighboring turns of DNA, creating a channel where neighboring grooves touch. The histones have tails, which can be seen in the new pictures to poke through the channel of alternate minor grooves.

The protruding tails are a feature of great interest to biologists because they signal the next level of structure beyond the nucleosomes. Some of the protruding histone tails make contact with the face of neighboring nucleosomes and probably determine how the nucleosomes stack together in clusters. These clusters, linked together by intermediate strands of DNA like beads on a string, can be seen under powerful microscopes when a chromosome, the highest level of DNA packaging, is chemically treated so as to uncondense its coils.

Whenever the cell needs to manufacture a new protein, it must retrieve a copy of the protein's specifications from the coiled stacks of its DNA library. The copying machinery, a floating blob of special proteins called a transcription complex, must somehow break down the higher order structure to get access to the DNA, and this probably requires some kind of interaction with the histone tails.

"What's really amazing," Dr. Richmond said, "is that there are so many people in the field of transcription who thought, 'We don't have to worry about the nucleosomes, they just get out of the way.'" Much of the process of activating a gene, in his view, lies in unraveling the nucleosome-based packaging that keeps the genes boxed in and silent.

Dr. Richmond said he took 18 years to obtain the high-resolution structure, in part because natural nucleosomes chopped out of chromosomes make irregular crystals that are unsuitable for X-ray analysis. The structure of DNA varies according to the information it is carrying, and the cell adds various bells and whistles to the histones after they are synthesized; and so, few natural nucleosomes are identical to their neighbors.

To obtain uniform crystals, Dr. Richmond learned that he had to genetically engineer his own histones and DNA. Dr. Rhodes, a former colleague at the Cambridge laboratory where Dr. Richmond started work on nucleosomes, described the new crystals as "a real tour de force of biochemistry."

At Zurich, Dr. Richmond's team had access to a powerful new X-ray machine in nearby Grenoble. X rays cannot be focused, but when zipping through crystals they will scatter into a pattern of spots. From the pattern, and other information, computers can act like a lens by reconstructing the picture the X rays would give if they could be focused. The basic picture shows the positions of the electrons that scattered the X rays. A graphics program can then infer the exact three-dimensional arrangement of the atoms and molecules in the crystal. For ease of viewing, the structure is usually drawn in schematic form, with the backbone of proteins or the helix of DNA represented by continuous ribbons, not the hundreds or thousands of atoms of which they are composed.

—NICHOLAS WADE, October 1997

A First Step in Putting Genes into Action: Bend the DNA

BY TRADITIONAL RECKONING, DNA is the benign dictator of the cell, the all-knowing molecule that dispenses commands to create enzymes, metabolize food or die a gracious death. But a series of new discoveries suggest that DNA is more like a certain type of politician, surrounded by a flock of protein handlers and advisors that must vigorously massage it, twist it and, on occasion, reinvent it before the grand blueprint of the body can make any sense at all.

Scientists have recently found an extraordinary class of proteins that serve almost exclusively as molecular musclemen, able to grab the cell's genetic material and, in a fraction of a second, kink the strands into hairpin curves. Then, just as quickly as they flexed the molecule of life, the proteins may jump off and allow the DNA regions to snap back to straightness.

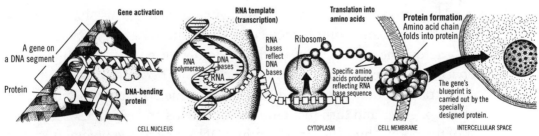

Patricia J. Wynne

Twists and Turns in Genetic Expression
The complex chemical reaction that leads from a dormant gene on the DNA molecule to an active protein that can perform tasks in the body is essential to life, and scientists now believe that the process often begins with a sharp bend to the DNA molecule. First, a protein attaches to the double helix and makes it crimp sharply. That bending brings together other molecules into an active switch, which can then start creating a chemical template that match the genes'. A strand of RNA is produced that then drifts out of the cell's nucleus into the cytoplasm, where it connects with a processing organelle called a ribosome. Like a laser scanner in a supermarket reading a bar code, the ribosome reads the RNA sequence and translates it into amino acids, the building blocks of a protein. The chain of amino acids is specially constructed according to the gene's instructions. After the product folds up and becomes a protein, it is potent enough to direct chemical and cell activity.

101

The act of bending DNA is turning out to play a critical role in controlling genes. In some cases it brings together far-flung bits of genetic information into a working command for the cell; at other times it allows teams of proteins to join forces at one well-bent spot and galvanize biochemical activity. By momentarily scrunching up parts of the chromosomes, the benders seem to influence the verve of the immune system, the sex of an infant, the scrimmages between a virus and its host. And the proteins add compelling new evidence to scientists' growing conviction that the complex architecture of DNA is at least as crucial to the behavior of genes as is the sequence of chemical letters of which the genes are composed.

As the most persuasive evidence of the importance of bending, scientists have recently discovered that a protein celebrated as the key to masculinity is a potent DNA bender. The protein, called testis-determining factor, was first identified in 1990 as the long-sought maleness signal, a molecule that somehow sets off a cascade of biochemical events and helps transform a fetus otherwise destined to become a female into a baby boy.

Researchers now have some idea how that factor accomplishes its epic task. Reporting in the journal *Cell,* Dr. Rudolf Grosschedl and his colleagues at the University of California at San Francisco said that the factor can manipulate DNA at many spots up and down the chromosomes, bending whatever straight region it homed in on by almost 90 degrees.

They propose that through a sequence of bends, this sharp bowing of DNA brings into contact molecules that otherwise are dispersed uselessly across the double helix. And once joined into squadrons by the go-between bender, the molecules are thought to become active switches, called transcription factors and enhancers, which are able to mobilize a battery of genes and turn them into new proteins and enzymes, the laborers of the body. Thus, the act of bending DNA may be the pivotal first step in switching on genes and getting anything substantive done during the growth of a fetus.

But the scientists warned that much remained to be learned about the bending protein, and about exactly which genes it helps flick on to initiate, for example, the fashioning of immature sex organs into working testes.

In the same report, the scientists described another protein, called lymphoid enhancer factor, that somehow influences the production of the T cells of the immune system. That protein is an even more emphatic DNA bender, crimping angles of 130 degrees into the chromosomes and, like the testis factor, presumably introducing distant transcription molecules to one another and thus igniting gene activity. In this case, the stimulated genes help create T cell receptors, proteins extruding from the surface of immune cells that can recognize nearly every foreign invader storming the body.

The bending of DNA is also involved in more nefarious events, including the ability of a virus to invade a host's chromosomes. Researchers studying the complex interplay between a bacterial cell and its parasitic virus, the phage, have discovered that the virus gets into the host chromosome by exploiting the bacterium's own DNA bending proteins, generating crimps in the DNA and then sneakily cutting and pasting its genetic information into the curled-over sequence.

Biologists believe that many human viruses, notably the one that causes AIDS, adopt a similar ruse, bending the DNA and then integrating permanently into the chromosomes. They suggest that by understanding normal DNA bending they may devise novel strategies to foil viruses before the deadly kinking and splicing occurs.

Other researchers, searching to understand exactly how chemotherapeutic drugs manage to destroy tumors, have discovered that at least one widely used drug, called cisplatin, seems to kill rapidly dividing cells by bending the cells' DNA into an abnormal configuration. That bending in turn attracts other proteins to the pleated areas, a molecular caucusing that ends up blocking DNA replication and killing the cell. Scientists hope that by learning the biochemical basis for a drug's success, they can design even more potent anticancer therapeutics in the future.

Above all, biologists said, the revelations about DNA bending emphasize the importance of the dynamic architecture of DNA to its performance. The emerging work on DNA benders and other proteins that tweak, sheathe, wiggle and whet the double helix demonstrate that DNA is in continuous motion and forever communicating with the throngs of molecules around it. That ceaseless activity and flexibility means that the chemical instructions encoded within the individual genes can be read in a magnificent variety of ways.

"I've been thinking about the architecture of DNA for most of my career, because it's so fascinating," said Dr. Lucy Shapiro, a developmental biologist at Stanford University. "And the superstructure of DNA is dynamic; it changes with time. By having a dynamic process, you add another level of control to your system."

As with so many of the greatest discoveries in the field of gene control, much of scientists' understanding of DNA bending comes from bacterial studies. Bacterial DNA is shorter and simpler than mammalian DNA, and all its parts and switches are designed for optimal performance, to permit the microbes to replicate as swiftly as possible. Scientists studying the amenable DNA soon realized that while DNA, when considered in its entirety, has many bends and curves in it—hence the image of a winding helix—short regions of it nevertheless are quite stiff.

"In essence, DNA is a rigid rod up until a certain length, when it becomes more pliable," said Dr. Steven D. Goodman, a senior staff fellow who works with a DNA bending pioneer, Dr. Howard Nash at the National Institutes of Health in Bethesda, Maryland. "Think of sticks. A long stick, like the bow of a bow and arrow, is bendable, but a small stick is virtually inflexible." On the scale of DNA, he said, a "small stick" is a sequence of several hundred subunits or less; the entire human DNA molecule runs for three billion subunits or bases.

But scientists knew that the tautness even in short stretches sometimes relaxed. They learned that when certain sequences of DNA happened to occur together, that region of the chromosome was more pliant and curved than other sections of the DNA of a similar size. For example, a repeated string of the DNA base called adenine would result in the chromosome's bending slightly at that point.

Those curving regions also proved to be the site of what Dr. Donald Crothers of Yale University, another leader in the field, called "interesting activity." Strings of adenine seemed to be located at parts of the chromosome where gene transcription began: where DNA would be written into another chemical form called RNA, the first step toward the creation of proteins. Somehow, inherent bending helped arouse genetic activity.

Other researchers discovered that when a virus infiltrated the bacterial DNA, it managed to bend the region sharply before stitching its genetic material into the host chromosome. That finding led to the isolation of a bacterial protein called integration host factor, or IHF, which can bend DNA by about 140 degrees, or almost back over on itself. It turns out that the bacterium normally uses the bending factor to mix together, or recombine, its genetic material during ordinary cell division. But the crafty phage virus has learned to use the IHF protein to fake the bacterium into thinking its DNA was bending simply for the sake of healthy recombination; instead, the virus slips its own genetic luggage into the wrinkled region.

Scientists subsequently identified a giant family of proteins related to the integration host factor, and they have been systematically investigating the molecular tribe to see which ones can bend DNA and to what end. Some of the proteins head for specific sequences of DNA and then sharply deform the region. Others are more profligate benders, deforming almost any patch of chromosome they alight upon.

Biologists are just beginning to determine why bending has a dramatic impact on DNA, but two overall themes are emerging. They have learned that the act of turning on any gene and translating it into a newborn protein is quite complex; nothing works by a mere on-off switch. Preceding a gene are strings of genetic sequences that allow the cell's protein manufacturing equipment to attach to the DNA and to work at a given speed and force. Those instructions often are quite spread out, over dozens to hundreds of bases, and scientists had long wondered how the codes could be brought together into a single legible how-to manual. Now they believe that the DNA is bent and furrowed to put the otherwise nonsensical words into their proper order.

"If you want a control region with a lot of information, and that information is strung out in a straight line, all the control elements can't communicate with each other," said Dr. Crothers. "So you have to fold the structure up."

At other times, DNA must be bent not to bring genetic sequences into proximity, but to permit contact between proteins that cling to the double helix like barnacles on a whale's back. Often these proteins need to meet their partners before they can perform their duties, which may be tending the DNA molecule, keeping it in good repair, replicating it before cell division, or even performing a task in the cell unrelated to the great molecule of life. In such cases, bending proteins will strong-arm the DNA to allow the proteins to touch, and the exalted double helix becomes nothing more than a kind of Rube Goldberg transportation device to get two proteins together.

"It's like a floppy disk for a computer," said Dr. Goodman. "Normally the floppy disk feeds instructions to the computer; it's software. DNA is mostly software for the cell. But if you were to use a floppy disk to, say, prop up the leg of your desk, then the software would for the moment be a bit of hardware."

Exactly how proteins deform DNA is not entirely certain, but scientists suspect that the DNA instantly wraps itself around the protein on contact, perhaps because there is some sort of affinity between protein and DNA.

"You have an essentially round protein contacting a linear DNA molecule," said Dr. Grosschedl. "To extend the contact between the two, the linear bends around the circular, introducing a deformation."

Researchers believe that, while bending proteins obviously are important to bacterial cells, the protein family is likely to be even more critical to so-called eukaryotic cells, the components of higher creatures like humans. That is because the DNA of higher creatures is wrapped up in protective packages of proteins, notably the histone proteins, which must be elbowed aside quite firmly to allow the genes within to burst to life. The bending proteins seem to compete with histones for the privilege of latching onto DNA, but whereas the histones pack it tightly away, the benders jostle it into action.

What is more, the intricacy of gene control in animal cells far exceeds that in bacteria. Scientists are still finding controls that control the controls in charge of the genes, amounting to what some have described as an overwhelming bestiary of molecules, enhancers, transcription factors and promoters. The ceaseless flexing of DNA may prove to be the easiest way for the cell to keep its restless fauna in line.

—NATALIE ANGIER, August 1992

Keys Emerge to Mystery of "Junk" DNA

IN RECENT TIMES, the twisted, viscous molecular celebrity called DNA has been described by any number of lofty metaphors. It is the book of life. The master molecule. The blueprint for a human being.

Yet to some researchers who consider the whole molecule, and not just the individual genes arrayed along its chemical coils, a few more homespun comparisons might better apply: DNA as your grandmother's attic, for example, or the best little flea market in town.

And still others say human DNA should be thought of as a sort of microscopic ecosystem, an invisible habitat teeming with competing bits of genetic material that often behave like benign yet selfish parasites, utterly indifferent to the needs of the human host cell in which they persist.

These analogies spring from recent explorations of the vast regions of the double helix that do not serve as recipes for creating the body's proteins, regions often given the pejorative if irresistible description "junk" DNA.

Of the three billion chemical building blocks, or bases, that make up human DNA, a mere 3 to 5 percent rate as coding regions—the genetic instructions for generating hormones, collagen, hemoglobin, endorphins, enzymes and all the rest of the body's proteinaceous workforce. That leaves 2,850,000-plus bases to account for, sentence upon page upon volume of genetic sequences that on first pass do not seem to say anything. Gibberish, filler, Styrofoam, junk, and all of it crammed into the core of nearly every cell of the body.

But as scientists are fond of pointing out, one person's junk is somebody else's treasure. Researchers are learning that much of this noncoding DNA must play essential roles in the performance of the genes embedded in it. They have determined that certain sequences once thought to be unnecessary and thus not subject to the same corrective forces that keep genes intact from one generation to the next in fact are highly conserved:

They have remained pretty much the same chemically over tens of thousands and in some cases millions of years of evolution, just as genes often do, which means this supposed junk must be indispensable to the organisms bearing it.

In some cases, the junk is thought to act as subtle enhancers of genes, turning their activity up from a murmur to a shout. In other cases, the junk tells the chromosomes what shape they are supposed to be as they are flexed and pleated into the nucleus of the cell.

Certain regions of junk may act as reservoirs of change, allowing the DNA to be more easily shuffled, mutated and rearranged into novel patterns that hasten evolution along. They are the curios in the attic that seem extraneous today, but that somebody later will stumble on, polish and haul back down to the living room for all to admire.

Still other noncoding stretches may be buffers against precipitous change, serving rather as flak jackets to absorb the impact of viruses and other genetic interlopers that infiltrate an animal's chromosomes. Without all the extra padding to absorb the blows, viruses or the bizarre genetic sequences that hop and skip from one part of the chromosome to another—mysterious genetic elements called transposons or jumping genes—might land smack in the middle of a crucial gene, disrupting its performance.

The new work appears to justify the claims of some proponents of the Human Genome Project, the huge federal effort to understand the entire complement of human DNA, that is, the genome. While financial pragmatists have counseled sticking with decoding the tiny regions of the genome that contain the 50,000 to 100,000 genes proper, those who wallow in junk insist that all three billion bases deserve attention. They suspect many of the most interesting insights into human evolution and large-scale genomic logic will come from looking at the abundant stuff around and between the genes.

"I don't believe in junk DNA," said Dr. Walter Gilbert of Harvard University, a preeminent theoretician of the human genome. "I've long believed that the attitude that all information is contained in the coding regions is very shortsighted, reflecting a protein chemist's bias of looking at DNA." Coding regions may make the proteins that are dear to a chemist's heart; but true biologists, he added, know that much of the exquisite control over these proteins is held offstage, nested within the noncoding junk.

Reporting in the *Proceedings of the National Academy of Sciences,* Dr. Roy J. Britten of the California Institute of Technology, who first described junk DNA 26 years ago, said that some of the most familiar junk in primate DNA has all the signposts of molecular raison d'être. These sequences, called Alu sequences, are short, repetitive strings of about 280 DNA bases apiece, which are scattered widely throughout the chromosomes of all primates, including humans. They have long been viewed as the meaningless remnants of an ancient impact event, the insertion of a virus-like bit of DNA into a protomonkey's chromosomes that was never tossed out because it did no harm. By this notion, the Alu sequences have been slothfully piggybacking along for the ride ever since, gradually and benignly replicating over the years into the half million or so copies now observed in primate DNA.

Dr. Britten proposes however, that whatever their origin, the Alu sequences have since been drafted into duty by the primate host, perhaps to serve as subtle modulators of the genes they are near. He said that the Alu sequences are too highly conserved to be explained away as useless molecular hobos. In addition, some of the most conserved parts of the sequences are just where one would expect them to be if they were to act as docking sites for the proteins that flick on genes or turn them up to high volume.

"It may be on the edge to claim that what's been considered the preeminent junk is under selective pressure and is probably carrying out some function," said Dr. Britten in a telephone interview. "But I take the general position that if there's something ubiquitous around, it will get used."

In the journal *Nature Genetics,* Dr. Ben F. Koop of the University of Victoria in British Columbia and Dr. Leroy Hood of the University of Washington in Seattle said they had compared huge stretches of human DNA with corresponding mouse DNA. They looked at 100,000 genetic bases that make up the gene responsible for the body's production of the T cell receptor, a critical part of the immune system. They compared those coding parts of the sequence that actually dictate the construction of the receptor, as well as the parts that lay in between—the so-called introns that are normally edited out during the multistep process of generating a protein as so many chemical *uhs* and *wells.*

Much to their surprise, they discovered that not only were the instructions for the receptor remarkably similar from mouse to man—an expected finding for a gene so vital to an animal's immune system—but so too were the supposedly trashy introns. Yet these introns are exactly the parts of the sequence that should drift about and change randomly over the 60 million years separating rodents and humans. Why bother keeping a tight leash on boorish genetic interruptions that presumably are not needed for the final performance of the receptor?

"When we find this sort of conservation, we have to think that even introns are involved in chromosome structure or organization, or some regulatory function," said Dr. Koop. "'Junk' to me is just a euphemism for 'I don't know.'"

However, he said, other introns and noncoding regions of the genome do differ greatly between rodents and humans, suggesting that in some cases no conservation is required. Quite the opposite: These areas could be the caches of mutability and evolutionary change, a safe testing ground where new genetic ideas may arise without deactivating existing genes. Eventually, the changes may be incorporated into an animal's coding region through deft chromosomal shuffling and a novel protein may be born.

"I'm of the school of thought that junk DNA is absolutely necessary in evolution and recombination," said Dr. J. Craig Venter of the Institute for Genomic Research now in Rockville, Maryland.

Dr. Venter pointed out that there is a good excuse for those who have long shrugged off junk as so much garbage. After all, some organisms function just fine without it. The smallpox virus, the *E. coli* bacterium and other microorganisms have only genes packed shoulder to shoulder, with no junk, or introns, in between. Even a couple of higher creatures, like the puffer fish, have very little noncoding DNA, which is why Dr. Sydney Brenner, a molecular geneticist at the Medical Research Council in Cambridge, England, has recommended using the puffer fish genome as a shortcut to understanding the chromosomes of all higher beasts.

As Dr. Britten and others have demonstrated, however, the vast majority of complex organisms have complex genomes, with a lot of nongene sequences. The two types of sequences, gene and nongene, can often be distinguished with relative ease. Genes tend to have a lot of diversity in their sequences; that is, the individual chemical bases of G, T, A and C are written out in fairly complex combinations. By contrast, junky sequences often are simpler and more redundant, consisting of the same letter or two repeated dozens of times. But this is only a rule of thumb, and scientists have often been misled by assuming a gene was junk just because it displayed a bit of monotony. Similarly, noncoding areas often exhibit considerable intricacy.

Scientists have also learned that there is a pattern to the various types of higher genomes. Most mammals, whether human, vole, cat or mole, have genomes about three billion bases long, indicating that junk and genes have remained in some sort of as yet mysterious state of equilibrium throughout the lengthy evolution of mammals. And again for reasons that are utterly unclear, plants often have even longer genomes than do mammals. The DNA of wheat, for example, measures about 16 billion bases long, while that of field lilies is about 100 billion bases long; and most of that extra DNA consists of lengthy, noncoding monologues.

Dr. Gilbert proposes that microorganisms like smallpox viruses and bacteria have tiny, tidy genomes out of necessity. They multiply rapidly and cannot afford to carry around anything not strictly necessary to immediate reproduction. But higher animals, with reproductive and life strategies a bit more involved than the fissioning in twain every 20 minutes, can afford the luxury of a more sophisticated genome, and at times make a virtue of its long-winded ways.

—NATALIE ANGIER, June 1994

Tracing a Genetic Disease
to Bits of Traveling DNA

SIFTING THROUGH THE CHROMOSOMES of people with an inherited nerve disease, researchers have made a strange discovery: Woven neatly into the human genome lie stretches of DNA once thought to belong exclusively to insects.

The finding does not mean that humans harbor an untapped potential to sprout wings or proboscises, or that people are any closer than previously thought to flies and mosquitoes. Nor are bugs waging gene warfare on humans.

The errant DNA is simply not the kind that codes for insect parts. It is, rather, an entity called a transposon, a wandering, free-agent scrap of DNA that can cut open genes and insert itself into places where it does not belong. The only information it carries are the instructions for making the enzyme it needs to cleave DNA.

Also called "jumping genes," transposons exist in virtually every known species, and have probably done so for millions of years. They are classified into groups with names like "pogo" and "gypsy," in appreciation of their ability to move around. Scientists have known about them and have argued about where they came from and what, if anything, they are good for since the 1940s.

The curious thing about the newly found one in people is that it is a member of a category known as mariner transposons, which until recently were thought to have invaded only the genes of insects. And although it is not the first mariner transposon to be found in a person, it is the first to be linked to a disease. The researchers who found it suggested that it and other transposons may contribute to genetic disorders by causing breaks at vital points in the chromosomes.

At the time they identified the mariner, the scientists were not even looking for transposons. Directed by Dr. James Lupski, at the Baylor College of Medicine in Houston, they were studying Charcot-Marie-Tooth syndrome, a disabling genetic disorder that causes nerves in the hands and feet to deteriorate. Named for its discoverers, the syndrome has nothing to do with teeth.

The results, published in the journal *Nature Genetics,* were the latest surprise in a decade of research that has repeatedly led Dr. Lupski down paths so tortuous that, he said with a laugh, even other scientists sometimes lose his trail.

In 1991, researchers in his laboratory discovered the genetic defect that causes the syndrome. The defect was not a mutation, or biochemical change, in the structure of a gene; nor was a gene missing. On the contrary, there was an extra copy of part of a chromosome, a duplication that left the chromosome slightly longer than normal. That type of defect had never been linked to a human illness before, although scientists did know of disorders like Down syndrome that were caused by an entire extra chromosome.

"We don't usually think of excess DNA as much of a problem," Dr. Lupski said. But his patients, disabled to varying degrees, provided evidence that it could be.

In humans, the genes reside on 23 pairs of chromosomes. Generally, a given trait is determined by two copies of a gene, one on each of a matched pair of chromosomes. But in Charcot-Marie-Tooth syndrome, the partial duplication in one chromosome in pair 17 gives patients three copies of a series of genes, perhaps 30 or more.

Dr. Lupski's group zeroed in one of those genes as the key to the syndrome. It contains the instructions for making a protein in myelin, the fatty substance that encases nerve fibers in the extremities. Myelin acts as an insulator for electrical signals. It breaks down in Charcot-Marie-Tooth syndrome, and, as a result, so does nerve transmission, which can lead to atrophy of the hands and feet.

The duplication was a tipoff to the origin of the syndrome. Geneticists have known for a long time that duplications result when there is a glitch in a normal process known as crossing over. That process, which occurs during the formation of sperm and eggs, allows the paired chromosomes to swap pieces of DNA with each other. To do that, the chromosomes arrange themselves so that like segments line up, and some of the segments trade places. The result, recombination, is a vital means in all species of stirring the genetic pot and creating diversity—and it is one reason that children can turn out so different from either of their parents.

But the swap can go wrong. Sometimes, one chromosome gives up a bigger piece of DNA than it gets from its partner. In that case, the crossing over is unequal, and one chromosome winds up with a duplication, while the other sustains a deletion. If either one gets into a sperm or egg that goes on to become a baby, the child may be born with a genetic disorder.

Dr. Lupski and his colleagues showed that unequal crossing over near the myelin gene on chromosome 17 could produce a duplication that would lead to Charcot-Marie-Tooth syndrome. If, instead, the chromosome with the deletion were passed on to the next generation, the result would be another nerve disorder, a milder one called HNPP, which stands for hereditary neuropathy with liability to pressure palsies.

The missing piece of the puzzle, though, was why unequal crossing over should happen so often at the same spot on chromosome 17. Charcot-Marie-Tooth and HNPP are not uncommon diseases, and the same defects kept turning up in the majority of patients.

Dr. Lupski and his team knew that the answer should lie in the DNA sequence of the duplicated segment of the chromosome, a region of some 1.5 million DNA subunits, or base pairs. Lawrence Reiter, a graduate student, set about mapping that segment in DNA from healthy people and those with the nerve diseases.

Mr. Reiter found several things that, taken together, helped explain what was happening. First, the segment included two smaller sections at either end that were repetitions of each other. The myelin gene lay between them. The repetitions would increase the chances that the two chromosomes would misalign during crossing over: The first repetition on one chromosome, for instance, might mistakenly line up with the second repetition on the other. The misalignment could lead to segments of unequal size being exchanged.

But misalignment alone could not explain the high rate of problems in the region; for crossovers to occur, the chromosomes must also break, and the scientists wondered where exactly the breaks were occurring, and what might be causing them.

"To our surprise," said Dr. Lupski, "we also found that in the majority of patients, the crossover happened in a very small region. We wanted to know what the sequence characteristics were that might make that region a hot spot."

To find out, they compared the sequence to ones that had already been studied by other researchers and logged into computer banks.

"Larry threw the sequence into the database," Dr. Lupski said. What came out were mariner transposons, located quite close to the hot spot. "He pulled out sixteen, seventeen, twenty different insect transposons," Dr. Lupski said. The closest matches were to ones found in green lacewings, a type of fly, and in the African mosquitoes that carry malaria.

"He presented evidence which was hard to refute," Dr. Lupski said. "But I certainly wanted additional confirmation, and we also had people who do this kind of thing for a living, matching sequences, look at it." They backed up Mr. Reiter. Clearly, the mariner transposon existed in all human genes—in healthy ones as well as in those from patients with the nerve disorders.

Its proximity to a genetic defect supported an idea held by many geneticists: that transposons, because of their ability to break and enter DNA, are potentially agents of genetic change, and of evolution. Sometimes known as "selfish DNA," they usually carry little information beyond what they need to cruise around the genome. That consists of the instructions for making a transposase, an enzyme that frees them to move around by cutting them loose from their host's DNA. Then, although they are not viruses, they can, like viruses, use their hosts' genes to make copies of themselves.

The transposon that the Baylor team found turned out to be inactive, so mutated that it could no longer produce any enzymes. But the team suggested that it might still cause chromosome breaks, by acting as a target for other enzymes that can cut DNA. Those enzymes, Mr. Reiter said, could be produced by other, active mariner transposons in the human genome. None have been found so far, but their existence has not been ruled out, either.

Dr. David Hartl, an evolutionary biologist at Harvard University who found the first mariner transposon, agreed with that theory. But, he added, although the mariner in this case appears to be harmful, transposons might also help bring about helpful genetic changes. "Maybe a mariner can play a role in regulating rates of recombination and perhaps generating duplications that are useful," he said.

Based on its similarity to the mariners found in insects, Dr. Hartl said, this one probably originated in an insect. "But we don't know whether it came directly from insects to humans," he said. "For all we know it went from insect to lizard to chicken to human." Probably, it was carried by a virus, but Dr. Hartl said he could not rule out Mr. Reiter's suggestion that the malaria mosquitoes may have had a role.

"We can't say it came from an insect," said Dr. Hugh Robertson, an entomologist and geneticist at the University of Illinois at Urbana-Champaign. Dr. Robertson has been studying mariners intensively for several years. "But we can say its closest relatives are in insects. But did it come from an insect, or jump into insects? We don't know what was really going on."

Of more concern to Dr. Lupski were the implications for patients of discovering the mariner. He expected it to lead to simpler diagnostic tests that would make it easier to detect both Charcot-Marie-Tooth syndrome and HNPP. In addition, the finding may shed light on other genetic disorders. Looking farther ahead, he said, knowledge of the transposon may one day help researchers to develop gene therapies. Transposons themselves might be used to carry treatment directly to the genes.

"The classic concept behind gene therapy that most people understand is simple," he said. "There's a broken gene, so I'll fix it by putting in a good one." That wouldn't work with Charcot-Marie. "You'd make it worse if you put in an extra copy, and there ain't nothing broken to fix," he said. "But now you know you should consider a treatment that would regulate the gene. This helps us get an understanding of how to direct our therapeutic approach."

—DENISE GRADY, March 1996

Scientist Reports First Cloning Ever of Adult Mammal

IN A FEAT that may be the one bit of genetic engineering that has been anticipated and dreaded more than any other, researchers in Britain reported that they have cloned an adult mammal for the first time.

The group, led by Dr. Ian Wilmut, a 52-year-old embryologist at the Roslin Institute in Edinburgh, created a lamb using DNA from an adult sheep. The achievement shocked leading researchers who had said it could not be done. The researchers had assumed that the DNA of adult cells would

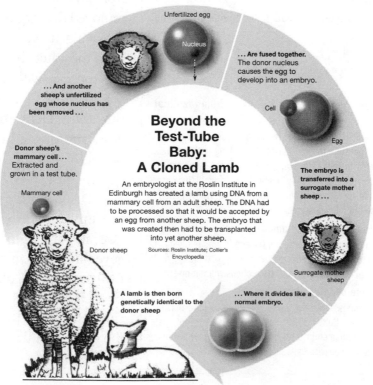

Unfertilized egg

Nucleus

... Are fused together. The donor nucleus causes the egg to develop into an embryo.

... And another sheep's unfertilized egg whose nucleus has been removed ...

Cell

Donor sheep's mammary cell ... Extracted and grown in a test tube.

Egg

Mammary cell

Beyond the Test-Tube Baby: A Cloned Lamb

An embryologist at the Roslin Institute in Edinburgh has created a lamb using DNA from a mammary cell from an adult sheep. The DNA had to be processed so that it would be accepted by an egg from another sheep. The embryo that was created then had to be transplanted into yet another sheep.

The embryo is transferred into a surrogate mother sheep ...

Donor sheep

Sources: Roslin Institute; Collier's Encyclopedia

Surrogate mother sheep

A lamb is then born genetically identical to the donor sheep

... Where it divides like a normal embryo.

117

N.Y. Times News Service

not act like the DNA formed when a sperm's genes first mingle with those of an egg.

A PRIMER

Questions About Genetic Possibilities

Could people be cloned?
Theoretically, yes. Sheep and humans are both mammals and the technology used to clone one species could theoretically be applied to the other.

Who are the parents of a clone?
In genetic terms, the parents of a clone would be the man and woman whose sperm and egg formed the person who, in turn, was cloned. In some states, the woman who bears a child is its legal mother.

Could a cell from any part of the body be used to make a clone?
Researchers do not know, but they believe that a kind of cell called a stem cell would be easiest to work with. These cells, which later give rise to a variety of other cells, occur all over the body, even in hair.

Could you clone the dead?
No. You have to fuse two living cells: the egg and the cell containing the DNA you want to replicate.

Could people who died and had their bodies frozen be cloned?
No. Their cells are dead.

Could a woman bear a clone of herself?
Theoretically, yes.

How would clones produced this way differ from identical twins?
Identical twins are born when a single fertilized egg splits in two, giving rise to two separate embryos. Scientists working with animals have long been able to split early-stage embryos in the laboratory to produce multiple births of genetically identical offspring. In this case, by contrast, the clone was made from an adult animal, whose genetic material was inserted into an egg cell.

Could you put the nucleus of a human cell into a sheep egg?
No. When the genetic material inserted in the egg begins to divide, it acts on instructions it receives from proteins in the egg. If the egg and the DNA are from different species, the instructions will not match.

Is it legal to clone people?
Britain, Spain, Denmark, Germany and Australia have laws barring human cloning, but the United States does not. Laws restricting use of federal funds on human embryo research might not apply because any cloning would be done with eggs, not embryos.

In theory, researchers said, such techniques could be used to take a cell from an adult human and use the DNA to create a genetically identical human—a time-delayed twin. That prospect raises the thorniest of ethical and philosophical questions.

Dr. Wilmut's experiment was simple, in retrospect. He took a mammary cell from an adult sheep and prepared its DNA so it would be accepted by an egg from another sheep. He then removed the egg's own DNA, replacing it with the DNA from the adult sheep by fusing the egg with the adult cell. The fused cells, carrying the adult DNA, began to grow and divide, just like a perfectly normal fertilized egg, to form an embryo.

Dr. Wilmut implanted the embryo into another ewe; in July 1996, the ewe gave birth to a lamb, named Dolly. Though Dolly seems perfectly normal, DNA tests show that she is the clone of the adult ewe that supplied her DNA.

"What this will mostly be used for is to produce more health care products," Dr. Wilmut told the Press Association of Britain, the Reuters news agency reported.

"It will enable us to study genetic diseases for which there is presently no cure and track down the mechanisms that are involved. The next step is to use the cells in culture in the lab and target genetic changes into that culture."

Simple though it may be, the experiment, reported in the British journal *Nature,* startled biologists and ethicists. Dr. Wilmut said in a telephone interview that he planned to breed Dolly to determine whether she was fertile. Dr. Wilmut said he was interested in the technique primarily as a tool in animal husbandry, but other scientists said it had opened doors to the unsettling prospect that humans could be cloned as well.

Dr. Lee Silver, a biology professor at Princeton University, said that the announcement had come just in time for him to revise his forthcoming book so the first chapter will no longer state that such cloning is impossible.

"It's unbelievable," Dr. Silver said. "It basically means that there are no limits. It means all of science fiction is true. They said it could never be done and now here it is, done before the year 2000."

Dr. Neal First, a professor of reproductive biology and animal biotechnology at the University of Wisconsin, who has been trying to clone cattle, said the ability to clone dairy cattle could have a bigger impact on the industry than the introduction of artificial insemination in the 1950s, a procedure that revolutionized dairy farming. Cloning could be used to make multiple copies of animals that are especially good at producing meat or milk or wool. It could allow the cloning of cows that are superproducers of milk, making 30,000 or even 40,000 pounds of milk a year. The average cow makes about 13,000 pounds of milk a year, he said.

Although researchers have created genetically identical animals by dividing embryos very early in their development, Dr. Silver said, no one had cloned an animal from an adult until now. Earlier experiments, with frogs, have become a stock story in high school biology, but the experiments never produced cloned adult frogs. The frogs developed only to the tadpole stage before dying.

It was even worse with mammals. Researchers could swap DNA from one fertilized egg to another, but they could go no farther. "They couldn't even put nuclei from late-stage mouse embryos into early mouse embryos," Dr. Silver said. The embryos failed to develop and died.

As a result, the researchers concluded that as cells developed, the proteins coating the DNA somehow masked all the important genes for embryo development. A skin cell may have all the genetic information that was present in the fertilized egg that produced the organism, for example, but almost all that information is pasted over. Now all the skin cell can do is be a skin cell.

Researchers could not even hope to strip off the proteins from an adult cell's DNA and replace them with proteins from an embryo's DNA. The DNA would shatter if anyone tried to strip it bare, Dr. Silver said.

In 1996, Dr. Wilmut showed that he could clone DNA from sheep embryo cells, but even that was not taken as proof that the animal itself could be cloned. It could just be that the embryo cells had DNA that was unusually conducive to cloning, many thought.

Dr. Wilmut, however, hit on a clever strategy. He did not bother with the proteins that coat DNA, and instead focused on getting the DNA from an adult cell into a stage in its normal cycle of replication where it could take up residence in an egg.

DNA in growing cells goes through what is known as the cell cycle: It prepares itself to divide, then replicates itself and splits in two as the cell itself divides. The problem with earlier cloning attempts, Dr. Wilmut said, was that the DNA from the donor had been out of synchrony with that of the recipient cell. The solution, he discovered, was, in effect, to put the DNA from the adult cell to sleep, making it quiescent by depriving the adult cell of nutrients. When he then fused it with an egg cell from another sheep—after removing the egg cell's DNA—the donor DNA took over as though it belonged there.

Dr. Wilmut said in the interview that the method could work for any animal and that he hoped to use it next to clone cattle. He said that he could use many types of cells from adults for cloning but that the easiest to use would be so-called stem cells, which give rise to a variety of other cells and are present throughout the body.

In his sheep experiment, he used mammary cells because a company that sponsored his work, PPL Therapeutics, is developing sheep that can be used to produce proteins that can be used as drugs in their milk, so it had sheep mammary cells readily available.

For Dr. Wilmut, the main interest of the experiment is to advance animal research. PPL, for example, wants to clone animals that can produce pharmacologically useful proteins, like the clotting factor needed by hemophiliacs. Scientists would grow cells in the laboratory, insert the genes for production of the desired protein, select those cells that most actively churned out the protein and use those cells to make cloned females. The cloned animals would produce immense amounts of the proteins in their milk, making the animals into living drug factories.

But that is only the beginning, Dr. Wilmut said. Researchers could use the same method to make animals with human diseases, like cystic fibrosis, and then test therapies on the cloned animals. Or they could use cloning to alter the proteins on the surfaces of pig organs, like the liver or heart, making the organs more like human organs. Then they could transplant those organs into humans.

Although Dr. Wilmut said he saw no intrinsic biological reason humans, too, could not be cloned, he dismissed the idea as being ethically unacceptable. Moreover, he said, it is illegal in Britain to clone people. "I would find it offensive" to clone a human being, Dr. Wilmut said, adding that he fervently hoped that no one would try it.

But others said that it was hard to imagine enforcing a ban on cloning people when cloning got more efficient. "I could see it going on surreptitiously," said Lori Andrews, a professor at Chicago-Kent College of Law, who specializes in reproductive issues. For example, Professor Andrews said, in the early days of in vitro fertilization, Australia banned that practice. "So scientists moved to Singapore" and offered the procedure, she said. "I can imagine new crimes," she added.

People might be cloned without their knowledge or consent. After all, all that would be needed would be some cells. If there is a market for a sperm bank selling semen from Nobel laureates, how much better would it be to bear a child that would actually be a clone of a great thinker or, perhaps, a great beauty or great athlete?

"The genie is out of the bottle," said Dr. Ronald Munson, a medical ethicist at the University of Missouri in St. Louis. "This technology is not, in principle, policeable."

Dr. Munson called the possibilities incredible. For example, could researchers devise ways to add just the DNA of an adult cell, without fusing two living cells? If so, might it be possible to clone the dead?

"I had an idea for a story once," Dr. Munson said, in which a scientist obtains a spot of blood from the cross on which Jesus was crucified. He then uses it to clone a man who is Jesus Christ—or perhaps cannot be.

On a more practical note, Dr. Munson mused over the strange twist that science has taken.

"There's something ironic" about this study, he said. "Here we have this incredible technical accomplishment, and what motivated it? The desire for more sheep milk of a certain type." It is, he said, "the theater of the absurd acted out by scientists."

In an interview with the Press Association, Britain's domestic news agency, Dr. Wilmut added: "We are aware that there is potential for misuse, and we have provided information to ethicists and the Human Embryology Authority. We believe that it is important that society decides how we want to use this technology and makes sure it prohibits what it wants to prohibit. It would be desperately sad if people started using this sort of technology with people."

—GINA KOLATA, February 1997

With No Other Dollys, Cloning Report Draws Critics

THE CREDIBILITY OF THE EXPERIMENT reporting the cloning of Dolly the sheep from the cell of an adult ewe is being sharply challenged by a leading biologist, and other eminent scientists agree that the experiment needs to be repeated before it can be accepted.

The skepticism erupted almost a year after the original report. The critics note that no other scientist has succeeded in cloning a mammal from an adult cell, although the birth of at least one calf cloned this way is said to be imminent. The cloned calves were generated from fetal cells, not those of an adult cow.

The challenge is in the form of a letter, published in the journal *Science,* by Dr. Norton D. Zinder, a microbiologist at Rockefeller University and a member of the National Academy of Sciences, and Dr. Vittorio Sgaramella of the University of Calabria in Italy.

In a response, the chief author of the cloning experiment, Dr. Ian Wilmut of the Roslin Institute in Scotland, dismissed the possibility of error but said that some extra tests suggested were under way and would be reported when completed.

Dr. Zinder's position is not that the cloning of Dolly never occurred, merely that so far there is not enough evidence to prove it.

In their letter, Dr. Zinder and Dr. Sgaramella note that Dr. Wilmut's cloning of an adult sheep was successful only one out of some 400 times and that in science, one success in 400 "is an anecdote, not a result."

They also criticize Dr. Wilmut's original report for failing to mention that the adult sheep from which Dolly was cloned had died several years earlier. Its absence prevented any direct comparison between Dolly and her donor, in particular the decisive test of a skin graft from one to the other. If true clones, each would have accepted the other's skin graft; if not, any graft would have been rejected.

Dolly was cloned from a vial of sheep breast cells that had been frozen away in Dr. Wilmut's freezer as part of another project. The critics say it is hard to know what other kinds of sheep cells may have been around in the vial but, since the sheep in question was pregnant at the time, the vial could have contained some of the fetal cells that circulate in the mother's body. If it was a fetal cell that generated the Dolly clone, the outcome would not be startling because Dr. Wilmut had already demonstrated that fetal cells could be cloned.

"The most heinous crime was not saving the parent," Dr. Zinder said.

"I just don't understand that. There could be nothing more exciting than seeing the two twins standing there," he said, referring to the fact that the cloned animals would be like identical twins although of different ages.

Dr. Wilmut said in a phone interview that he had not intended to clone an adult cell when he started the experiment, which originally dealt with fetal cells. Midway through that experiment, he said, he decided to try cloning an adult cell. Instead of working with a live animal, he said he used a cell line that was already available.

Dr. Zinder's criticisms have resonated among other biologists, in part because of a seminar Dr. Wilmut gave at the Massachusetts Institute of Technology. The seminar was attended by many of Boston's leading biologists, several of whom were surprised to hear him say he did not intend to repeat the experiment.

It is an article of faith among scientists that an experiment should be replicated in one's own laboratory in case of an error the first time around, as often happens. Also, it is usual to follow up an important result with more experiments.

Dr. Wilmut confirmed that he did not intend to replicate his experiment.

"I don't perceive a need," he said. "The principle is established. Repeating experiments is boring and unimaginative."

He said he expected his results to be repeated by many other laboratories.

So far three laboratories have tried and failed to repeat Dr. Wilmut's experiment, but others are still trying and are confident of ultimate success, according to an article in *Science*.

Other scientists disagree with Dr. Wilmut's view that one need not replicate one's own results.

"If I have one success in 215 tries and I don't try to duplicate it, then I'm concerned it's not science," said Dr. Philip Sharp, a Nobel Prize–winning biologist at MIT who attended the seminar.

Dr. Sharp described Dr. Wilmut as a "reputable scientist who has worked long and hard in the field." But, he added, "if you have only one success in an experiment which is so widely recognized from a public and now political perspective, then I am really leery."

Dr. Bruce Alberts, president of the National Academy of Sciences, said he could not comment on Dr. Zinder's letter. "However," Dr. Alberts added, "it is fair to point out that scientists generally require that a new result be repeated in at least one independent laboratory before they are ready to accept it unambiguously."

The issue of human cloning has received considerable public attention since Dr. Wilmut's announcement in February 1997. Politicians from President Clinton, in his State of the Union Message in January 1998, to President Jacques Chirac of France have called for a ban on cloning people.

Dr. Zinder said he wrote to *Science* because of his feeling that if there was going to be such an intense public debate over cloning, it would be useful to make sure the underlying facts were true.

"The whole ambiance of this story was that nobody ever questioned whether or not this was true, and that included the hundred-odd page report of the President's Commission on Bioethics," Dr. Zinder said. "Wilmut sat before Congress with Varmus sitting next to him and nobody ever said, 'Is this possibly not true?'" Dr. Harold Varmus is director of the National Institutes of Health.

For biologists, the cloning of animals from adult cells has been a long-standing barrier, but one founded more on practical than theoretical obstacles. A full-fledged carrot can be cloned from any cell of an adult carrot, whether a root cell or a leaf cell. So it would not in principle be surprising if the same were true of animals.

But the most serious effort to clone animals, an elaborate series of experiments on frogs in the 1960s, showed that cells from tadpoles could be cloned but that viable offspring could almost never be obtained from the cells of adult frogs.

The inference was that as an animal progresses from egg to embryo to adult, its cells become increasingly committed to specific fates. A skin cell, say, somehow switches off all its genes except those needed to operate a skin cell. The off switches can be reset in a tadpole's cells to generate a whole frog, but not in a grown frog's cells.

Efforts to clone even fetal cells in mammals failed for many years, and some biologists began to doubt that cloning would ever be possible from specialized adult cells, also known as differentiated cells.

"The cloning of Elvis Presley from fully differentiated cells is not something we should count on," declares a textbook on developmental biology. But biologists had been edging ever closer to that goal, or at least its general vicinity, as their techniques improved. By 1986 lambs had been cloned from very early embryos, and Dr. Wilmut in 1996 showed how lambs could be cloned from a later embryo's cells.

His report in February 1997 that he had made the leap to cloning adult cells was more credible because of his previous work in the field. To prove that Dolly's and her donor's DNA were identical, however, Dr. Wilmut used only a single test, the same DNA typing technique that is now accepted in courts. Dr. Zinder said this test was strong but not conclusive, because domestic animals are often highly inbred, meaning that the DNA from two different sheep could quite possibly be identical. Dr. Wilmut said the assertion that sheep are highly inbred was a "silly exaggeration."

The doyen of animal cloning, Dr. John B. Gurdon of Cambridge University in England, who performed the frog cloning experiments of the 1960s, said he believed that Dr. Wilmut's experiment showed an animal could be cloned from an adult, but whether the cell was differentiated was "a point that has not been shown by Dolly the sheep." The frog evidence was more compelling, Dr. Gurdon said.

Dr. Wilmut has no doubt that other laboratories will be able to repeat his success.

"Dolly is real," he said. "In the next few months you'll see positive results coming in from other labs and people will accept it for what it is."

—NICHOLAS WADE, January 1998

In Big Advance, Cloning Creates Dozens of Mice

SCIENTISTS FROM HAWAII report that they had made dozens of adult mouse clones and even cloned some of those clones. The announcement, coming after months of rumors, still stunned biologists when they heard the details. It means, they say, that advances in cloning are coming faster than even the most confident scientists had imagined.

The research team took just months to churn out clones of adult mice, following the announcement in 1997 that the first clone of an adult animal, Dolly the sheep, was created in Scotland.

In *Nature,* Dr. Ryuzo Yanagimachi, a 69-year-old biologist at the University of Hawaii, and his postdoctoral student, Dr. Teruhiko Wakayama, report on the first 22 mice that are clones, seven of which are clones of clones. They say they have made a total of more than 50 mouse clones.

The feat, scientists said, meant that cloning an adult animal like Dolly is not a fluke as some have suggested. And with mice, they said, researchers can study and perfect cloning in an easily available and familiar lab animal.

"Wow," said Dr. Barry Zirkin, who is the head of the division of reproductive biology at Johns Hopkins University in Baltimore. "This is going to be Dolly multiplied by twenty-two." In cloning mice, Dr. Yanagimachi and Dr. Wakayama defied conventional wisdom among biologists, who had long contended that perhaps mice would be the one mammal that was impossible to clone because of the extraordinarily rapid development of the mouse embryo just after fertilization.

Dr. Lee Silver, a mouse geneticist and reproductive biologist at Princeton University, described the speed at which the cloning progressed as breathtaking.

"It's absolutely incredible," Dr. Silver said. "He did all this in the past year."

The implications of the work were clear, Dr. Silver added. "Absolutely," he said, "we're going to have cloning of humans."

"If we follow scientific protocol," Dr. Silver said, "it could take five to ten years before in vitro fertilization clinics add human cloning to their repertoires." The method would first have to be perfected in mice and then monkeys, he said.

Dr. Zirkin said that human cloning "needs to be discussed," but he added that the main significance of Dr. Yanagimachi's announcement lies in the opportunities it creates for scientists to explore questions about basic reproductive biology.

"One of those questions, but only one," Dr. Zirkin said, "has to do with the cloning of humans."

The concept of genetically duplicating a person or animal by using the DNA of a single cell has captivated scientists and the public for decades.

But until the birth of Dolly was announced, most scientists had given up on the idea of cloning adults, although they generally agreed on the potential for cloning fetal cells. Even though every cell in the body has the same genetic material, cells of adults have reached their final state of development and never budge from it. Heart cells do not become liver cells, lung cells do not become brain cells.

To clone an adult, the genetic material from one of these cells essentially must go backwards in time and enter the state it was in when sperm first fertilized egg.

On February 23, 1997, the world learned that Dr. Ian Wilmut of the Roslin Institute and Dr. Keith Campbell of PPL Therapeutics in Roslin, Scotland, had created Dolly by cloning from an udder cell of a six-year-old ewe.

Critics soon emerged, led by Dr. Norton D. Zinder, a microbiologist at Rockefeller University in New York. Dr. Zinder questioned Dr. Wilmut's genetic evidence that Dolly was a clone.

Dr. Wilmut responded to Dr. Zinder in two papers also published in *Nature*. Dr. Wilmut's group and an independent group that included the inventor of DNA fingerprinting, Dr. Alec J. Jeffreys of the University of Leicester, compared DNA from the original piece of udder—frozen at the Hannah Research Institute more than 100 miles from the Roslin Institute—with both the DNA of Dolly and the DNA of the udder cells that were frozen in a test tube at the Roslin Institute and used to create Dolly. All the DNA sequences were identical, the researchers reported.

But Dr. Zinder, in a telephone interview, said he still was not convinced. The investigators, he said, did not prove that the piece of udder that they had stored at the Hannah Research Institute really was the source of the test tube of cells used to create Dolly.

"They didn't keep proper records," Dr. Zinder said. "We don't know what that chunk of tissue was."

Dr. Zinder also questioned Dr. Yanagimachi's work. "I'm no mouse expert, but it looks to me like it's very very shaky," he said.

Those who are mouse experts did not share Dr. Zinder's doubts about the cloning.

"I am convinced," said Dr. John J. Eppig, an expert in mouse embryo development at the Jackson Laboratory in Bar Harbor, Maine.

Dr. Richard Schultz, an expert in early mouse embryo development at the University of Pennsylvania, offered this assessment: "The significance of this one is that it's incontrovertible. A superb experiment."

In his cloning experiment, begun just a year ago, Dr. Yanagimachi used one of Dr. Wilmut's key ideas, but varied his method.

Dr. Wilmut proposed that the secret to cloning was to put a cell into a resting state, so that it was not dividing, before using it to clone. He did this by starving the udder cells so that they went into a state of hibernation. Then he slipped one of those cells into a sheep's egg whose own genetic material had been removed and gave the egg a shot of electricity to start the development.

In contrast, Dr. Yanagimachi and Dr. Wakayama started with three types of cells that were already in a resting state: cumulus cells, which cling to eggs like a thick smear of caviar; Sertoli cells, which are the male equivalent of cumulus cells, and brain cells. That experiment indicated that cumulus cells would be easiest to clone, so the scientists used them exclusively.

Dr. Yanagimachi and his colleagues injected the cumulus cell's genetic material into mouse eggs whose own DNA had been removed. They waited six hours to give the egg a chance to reprogram the cumulus cell's DNA and then chemically prodded the egg to start dividing. The process of reprogramming remains a mystery to the scientists.

All of the mouse clones were female.

Dr. Yanagimachi and his colleagues verified with genetic tests that their mice were indeed clones. Also, in one experiment, they used coffee-colored mice for the cumulus cells, black mice for the eggs, and white albino mice as surrogate mothers. As predicted, the clones were coffee colored.

Despite the enthusiastic reception for the results of Dr. Yanagimachi's experiment among reproductive biologists, he had great difficulty publishing his paper. On October 5, 1997, after cloning the first four mice, he submitted a paper to the journal *Science,* which rejected it without peer review, telling him, he said, that it was "not of general interest." Diane Dondershine, a spokeswoman for *Science,* said that the journal's policy is not to comment on papers that were submitted.

Dr. Yanagimachi then sent his paper to *Nature,* which forwarded it to two reviewers, one of whom asked repeatedly for additional proof. In March 1998, the journal sent the paper to two more reviewers before finally accepting it in June, Dr. Yanagimachi said.

In the meantime, the scientists kept cloning.

The process has become so straightforward, Dr. Wakayama said in an interview, that he has now made more than 50 clones and could clone every day with no difficulty and turn out baby mice after the normal 20-day gestation period.

And lest anyone doubt that the clones exist, the scientists carried a cage of mice to New York to show them off, including a coffee-colored adult mouse that was cloned, the clone created from that mouse, the black mouse that provided an egg for the cloning, and the white mouse that was the surrogate mother.

Already, a venture capitalist in Hawaii is setting up a consortium of companies and academic scientists to make the cloning of adult animals a commercial reality within a few years. But the investor, Laith Reynolds, chief executive of Probio America, said, "We have no interest in cloning humans."

"Besides being the politically correct answer," Mr. Reynolds said, "we can't see any business in it."

Now, scientists say, the opportunity is here to figure out how cloning works, how to make it work better, and how to apply it.

For example, Dr. George Seidel, a cloning expert from Colorado State University in Fort Collins, said that scientists wanted to understand how an egg reverses the program of a cell's DNA. Once they determine that, they might eventually be able to turn one cell into another.

"Let's say my pancreas was being destroyed for one reason or another," Dr. Seidel said. "If there are some cells in my body that can make a whole individual, they certainly can make a pancreas. It's pretty important, pretty potent stuff from a long-term standpoint."

And then there is the question of actually cloning people.

Human cloning, Dr. Seidel said, today "is clearly more imminent."

—GINA KOLATA, July 1998

5

BUILDING AN
ORGANISM

In Greek mythology the goddess Athena was born as an adult: She sprang, fully clothed and armed, from the head of her father, Zeus. Ordinary mortals are created in a rather more miraculous way. The single cell which is the fertilized egg grows and divides into a commonwealth of ten thousand trillion cells, each playing its allotted role in the organism.

Biologists have made a strong start in understanding how the genetic program of a multicellular organism unfolds. Right at the egg stage, genes are switched on so as to give the egg a polarity in two dimensions—this side is up, this side is down, front end here, back end there.

As the egg cell divides into two cells, then four, eight, sixteen and so forth, the pattern laid down in the egg is retained and made more intricate. Cells form into three main layers, one to make the skin, one to make the gut, and one to make everything in between. The cells in each layer of the developing embryo are tagged with a particular set of active genes that define their nature and position.

The organizational complexity of the embryo is now much greater than that of the egg. Its cells begin to differentiate, or commit themselves to specific fates, like being a brain cell or a hormone-secreting cell in the pancreas gland. At least 250 different types of human cells are now recognized.

Every ordinary cell in the body possesses the same set of genes, so the differences between cells must arise because a special subset of genes is switched on in each kind of cell.

The way an embryo develops depends heavily on cell signaling. The protein signal secreted by one cell induces its neighbors to assume certain roles, and they in turn influence their neighbors. The developmental program seems thus to consist of a series of genetic instructions for getting from egg to adult, not a blueprint of the final result.

The following articles describe new findings at several stages of the developmental program, as well as explanations at the genetic level for some of the adult's physiology, such as the sense of smell and the establishment of hair-growth patterns.

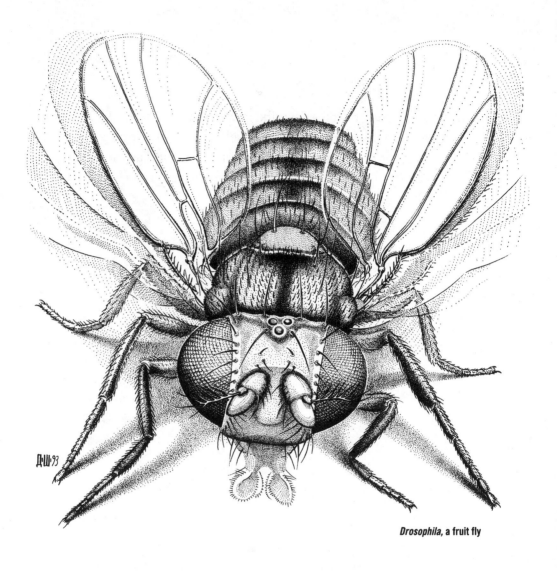

Drosophila, a fruit fly

C. elegans

Mouse

Human

Dimitry Schidlovsky

Heads or Tails?
How Embryos Get It Right

THE EARLY DECISIONS are always the hardest. Should I sleep a little longer, or piously go running? Should I get out of bed at all? Which shampoo will leave my hair the least flat? Read the paper and be depressed, or skip the paper and risk sounding stupid later on?

Yet as difficult as the average adult morning may be, it is a lazy Sunday compared with the dawn of life for embryos. Theirs is the ultimate wake-up call, for the first thing that any embryo must figure out before it rouses itself and gets to work is, which way is up? Whether destined to become a fly or a worm or a person or a tulip, the embryo must learn its top from its bottom, its front from its rear. Scientists call this initial stage of development "establishing early polarity," which essentially means the embryo must make heads or tails of itself.

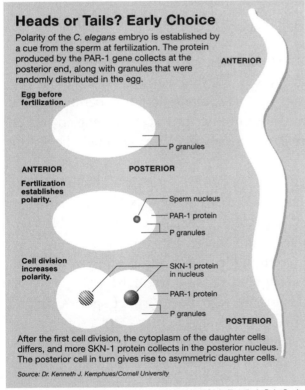

Heads or Tails? Early Choice

Polarity of the *C. elegans* embryo is established by a cue from the sperm at fertilization. The protein produced by the PAR-1 gene collects at the posterior end, along with granules that were randomly distributed in the egg.

ANTERIOR

Egg before fertilization.

P granules

ANTERIOR POSTERIOR

Fertilization establishes polarity.

Sperm nucleus
PAR-1 protein
P granules

Cell division increases polarity.

SKN-1 protein in nucleus
PAR-1 protein
P granules

POSTERIOR

After the first cell division, the cytoplasm of the daughter cells differs, and more SKN-1 protein collects in the posterior nucleus. The posterior cell in turn gives rise to asymmetric daughter cells.

Source: Dr. Kenneth J. Kemphues/Cornell University

N.Y. Times News Service/Illustration by Baden Copeland

137

If the early embryo is thought of as a beach ball, monotonous, symmetrical and directionless, then the initiating step of development is to break the symmetry, disrupt the conformity and distort ball into pear, as it were.

Paradoxically, new evidence suggests striking differences in the path of embryonic growth in different species at the very beginning of development, and similarities later on.

In some species, early polarity is set in the egg, when one pole of the cell begins to look and behave differently from the other pole. That initial difference is then amplified as embryonic growth proceeds. In other species, polarization comes a bit later, after the fertilized egg has split a few times into a cluster of identical cells. Only then does one cell "learn" that it is the top, and the one below it is the bottom.

Whenever it happens, polarity is the inaugural stage of pattern development, the first step in specifying the shape and form of a creature. It is a kind of graph, with all the compass points in place, on which the art of an organism can be sketched.

Scientists have made considerable progress in understanding this most embryonic stage of embryonic growth. They are sorting out the molecular and cellular signals that help set up early polarity in a number of different organisms. Many of the findings were described at a meeting in Boston under the auspices of the journal *Nature,* called "Patterns of Life: The Nature of Biological Development."

The meeting was packed with world-class scientists, including two Nobel laureates, Christiane Nusslein-Volhard of the Max Planck Institute in Tubingen, Germany, and Eric Wieschaus of Princeton University in New Jersey. Also in attendance were many graduate scientists and scientists-in-training, who seem not to be deterred from their plans and passion by the glum prospect of dwindling financial support for basic research.

By comparing and contrasting the signals in distinct species, among them fruit flies, nematode worms called *C. elegans,* frogs and a small, easily manipulated plant called *Arabidopsis,* researchers are learning both the degree of commonality among all the creatures of the Earth, and the extent to which they differ.

Beyond discussing the latest research on early polarity, the scientists described later stages of development, including the commitment of premature cells to a particular fate in one organ or another, and the eventual appearance of limbs and appendages—this last a favorite topic of developmental biologists.

As it turns out, the mechanisms used for establishing the earliest patterns in an embryo are far less similar than those relied on later in development, say, when the limbs or the nervous system are being formed.

In other words, the signals that the fruit fly uses to know its front end from its back turn out to be distinct from those that the nematode relies on; and both appear to be quite different from the polarity signals at work in the embryogenesis of mammals like humans.

For example, in the fruit fly, all the action in laying out the basic body plan occurs during the development of the egg, before fertilization even occurs. The signals for polarity are bestowed by the mother and are arrayed in the yolk-like cytoplasm of the egg. When the sperm comes along and contributes its genetic material, everything is set to go as though spring-loaded. Given all the positional information that must be distributed throughout the egg, it takes a female fruit fly a week to make an egg, said Dr. Daniel St. Johnston of the Wellcome/CRC Institute in Cambridge, England. But once that well-appointed egg is fertilized, he said, "it goes from a single cell to a larva in only a day." A larva, it might be added, that can crawl off perfectly well on its own, as a delightful little maggot.

By comparison, scientists have recently discovered that the *C. elegans* worm relies on sperm for giving it direction in life. In the worm, the sperm's point of entry into the egg appears to provide the first positional information that distinguishes front from back in the embryo. A similar sperm-driven mechanism is thought to establish body orientation in comparatively simple vertebrates like frogs, though not, to all appearances in higher vertebrates like mammals.

Once an embryo is growing and knows its left from its right, it relies for its elaboration of parts on sets of essential genes that are remarkably similar from worms and flies up through mammals. One much emphasized theme in developmental biology over the last several years has been the astonishing conservation of mechanism: The genes that help make eyes in flies are similar to the genes that make eyes in mice or humans, even though a fly eye looks completely different from that of a vertebrate.

So a seeming paradox arises. From a molecular standpoint, embryos are most dissimilar when they are at the one- or few-cell stage and are practically indistinguishable one species from the next; and they are the most similar when they start growing their brains or extremities and start looking like identifiable, distinctive beings.

But scientists admit that they cannot yet reach any conclusions about any phase of development, since their knowledge remains sketchy. And though the pace of discovery is so fast that it takes a strong pair of lungs for researchers to keep up, they are also finding that the more they learn, the more tangled embryogenesis appears. "Setting up polarity is very complex," Dr. St. Johnston told participants at the conference. "There are many more steps than we originally thought."

It is also a far more dynamic process than originally thought. The egg and embryo at their initial stages are seething sacks of activity. One scientist has taken a time-lapse movie of egg formation, or oogenesis, in the fruit fly. "The whole oocyte looks like a washing machine," Dr. St. Johnston said. "Basically all of the yolk is going round and round and round."

Much of scientists' understanding of early development comes from the analysis of fruit flies, or *Drosophila,* the king organism of biology's experimental vermin. The *Drosophila* egg is a magnificent cell, in which 10 different molecular signals are distributed in three spots to lend a gridwork pattern to the embryo that will eventually develop within. The molecular signals are so-called maternal RNAs, messages made from the mother's DNA that will eventually be translated into proteins in the egg cell to let the embryo know that one side is the anterior, or head side, the other the posterior side, and then to know its back, dorsal end, from its front, ventral end. Dr. St. Johnston said that much of the work in distributing these RNAs throughout the cytoplasm now appeared to be done by the so-called microtubules, which serve at once as the bones and railroad tracks of the egg cell.

In setting up the compass points of the egg, the microtubules exchange signals with follicle cells that surround the egg. Through that interchange, the very skeleton of the egg changes shape, to distribute maternal messages to their proper places. An RNA called bicoid ends up at the anterior end and a message called oskar defines the posterior end. (Most *Drosophila* gene names have their origin in clever references, used to describe the appearance or behavior of flies carrying a mutant version of the gene under study; for example, oskar was named for a mutation that resulted in a fly with a dwarfish body, hence the allusion to the antihero of Günter Grass's novel *The Tin Drum*.)

Once this anterior-to-posterior axis is defined, the microtubules are again rearranged, and the yolk churns with activity. The entire cell nucleus—the pouch with all the genetic information—is nudged from the posterior toward the anterior. In moving the nucleus, the microtubules also transport a third maternal message, called gurken, from the posterior to a point toward the front pole. Gurken RNA is a key maker of the second body axis, the one that says, here is the dorsal side where wings will someday grow, and here the ventral, where legs may soon sprout forth.

It is as though the struts, beams and walls of a building were to bend and flop around in order to convey a person from one story to another.

"It's quite possible that the organization of the entire cytoskeleton is dedicated to getting these various signals to their proper place," Dr. St. Johnston said.

A restless trafficking in parts is the hallmark of building an embryo, whatever the species. In the translucent, millimeter-long nematode, the egg appears to be ripe for polarization, but, unlike the fruit fly, it needs a signal from the outside to break its intrinsic monotony. "In *C. elegans* and some other species, the egg is primed and ready to go, but it needs an external signal to set up the initial asymmetry," said Dr. James R. Priess of the Fred Hutchinson Cancer Research Center in Seattle. "The point where the sperm enters the egg sets up a chain of events that is poorly understood."

However fuzzy the mechanism, the asymmetry is starkly evident almost immediately after fertilization, said Dr. Kenneth J. Kemphues of Cornell University in Ithaca, New York. Little bundles of proteins called P granules, which are initially spread uniformly throughout the egg cytoplasm, begin to congregate at one end of the yolk. As a result, when the fertilized egg divides in two, one daughter cell gets all the P granules, and the other daughter gets none. This distinctly inequitable distribution of protein particles is accompanied by strikingly different behavior on the part of the daughter cells, with one cell further dividing to become the worm's front, or anterior, and the other becoming the back end.

Nematode researchers are now parsing out what that P granule and other overt signs of asymmetry really mean when it comes to the much more elusive behavior of genes within the daughter cells. Looking at the two-cell worm embryo, Dr. Priess and his coworkers have determined that a gene called skin-1, for example, is quite active in the posterior daughter cell, but almost silent in the other, front-end cell. Importantly, skin-1 is a so-called transcription factor, which means it can switch on or off a whole battery of other genes and thus exert a powerful effect.

What these findings will mean to the understanding of the early life of humans is unclear. In a human or other mammalian embryo, polarity takes some time to develop. Through many stages of initial cell division, there are no apparent discrepancies or asymmetries between one cell and another. None seems to know its position or eventual fate, and all for a while are equally capable of giving rise to a complete animal. For now, the moment when the human grid of life is drawn remains as dark as ink.

—NATALIE ANGIER, November 1995

Making an Embryo:
Biologists Find Keys to Body Plan

AS ANY ARDENT VIEWER of the *Star Trek* television series will attest, the great majority of alien creatures portrayed, no matter how theoretically distant their origins, look terribly familiar. They have bodies separated into two basic regions, head and trunk. They have eyes, ears, mouths and snouts, albeit of varying puttied-up shapes and dimensions. They have arms and legs in tidy symmetrical pairs.

It is as though the producers of the show believe in a Platonic ideal of a body plan, a way of putting an organism together that has such firm trans-galactic logic behind it that the scheme has evolved independently, time and planet again.

But lest the creators of the series be accused of intellectual timidity or laziness, it turns out that nature, too, believes in the all-purpose, ever reusable blueprint for building mobile bodies. Researchers probing the earliest events in the transformation of a single fertilized egg into a breathing, sentient, multicellular being have discovered a class of genes that could well be the molecular signature of beasthood, the key to the kingdom Animalia.

The genes, called the Hox genes, are among the potentates of animal development, working in the first few days of embryonic growth to lay out the basic structure and orientation of a body—where the head will be, where the limbs, where the digits, where the chest, where the organs sheltered within. Of great importance, Hox genes speak in many tongues: Scientists first discovered the genes in fruit flies, but they have since detected signs of Hox activity in the early embryos of mice, worms, fish, chickens, grasshoppers, humans, frogs, cows, indeed in every creature studied to date.

The precise number of Hox genes found within an animal's cells differs significantly from vertebrates like humans, who harbor 38 such genes, and

Fertilized
egg

Small number
of identical
cells

Blastula
Hundreds of cells, all identical

Hox genes within each cell nucleus

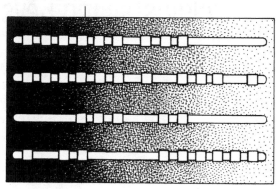

**Morphogenesis: Establishing the structures
and cell types for animal life**

The animal's body structure begins as a tiny
ball of undifferentiated cells. Scientists know
next to nothing about the earliest genetic signals, but have discovered the signals that
direct cells to begin to form as different organs
and tissues, the Hox genes. In mice, differentiation begins 7.5 days after conception and is
completed in 2.5 or 3 days.

Cluster of Hox genes   ■ Genes active and signaling

Blastula cells   Nucleus

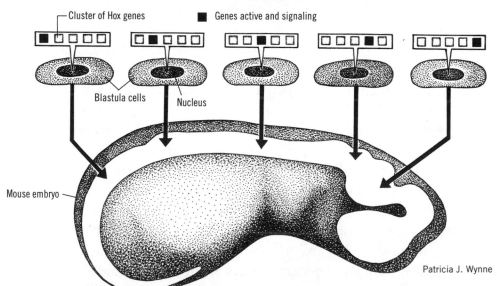

Mouse embryo

Patricia J. Wynne

Geography is destiny

Cells are assigned "addresses" by
Hox genes in an orderly sequence,
from anterior (head) to posterior (tail).
The first organ to be patterned is the
hindbrain. The process is completed
with the patterning of the genitals.

What activates the Hox genes?

How the Hox genes are activated is
still unknown, although scientists
believe it is linked to steroids, retinoic
acids, or other factors that are present in varying concentrations in the
embryo. Some genes would be activated in the presence of high concentrations of these chemicals, while
others are activated when the concentrations are low.

Once cells are assembled at their new
addresses, they begin to form organ
buds, skeletal material and all the
other constituents of the animal's
systems.

invertebrates like fruit flies, which make do with just eight; but the fact that species separated by 600 million years of evolution nevertheless rely on the same class of genes to orchestrate embryonic growth suggests that the molecules work too splendidly to demand much tinkering or reinvention.

"At each stage of our research we've had to suspend disbelief," said Dr. Matthew P. Scott, a professor of developmental biology and genetics at Stanford University School of Medicine, who has long studied *Drosophila*, or fruit fly, development. "But it's become very clear that to an astonishing extent all of us in developmental biology, no matter what our particular model system, are working on the same animal. There are enormous differences, but the amazing thing is how many components are shared."

So primary do Hox genes seem to be to the design of the animal body that researchers from Oxford University in Britain proposed in the journal *Nature* that the genes be considered the ultimate test of animalness, a potentially more precise definition than such traditional measures as the ability to move independently or to respond to stimulation. They point out that the genes are not found in the cells of plants, fungi or slime molds, and that therefore the DNA of any species whose phylogenetic status is in doubt— for example, sponges and some groups of protozoa—should be examined for evidence of Hox genes.

The Hox genes are not the sole players in the grand epic of animal development, and other molecules like steroid hormones, a derivative of vitamin A called retinoic acid and a host of growth factors also participate in the extravaganza of embryogenesis. But biologists have recently made great progress in cracking the puzzle of the Hox genes. They have created transgenic mice that are outfitted with telltale mutations in their Hox genes, and they have artfully rearranged the influence of the genes on the budding wing of a chick embryo.

Through such experiments, they have learned that Hox genes operate as master switches, producing proteins called transcription factors that clamp onto the chromosomes and turn on wave upon wave of subordinate genes, thereby amplifying a modest initiating signal into great heaves of biochemical activity.

Scientists have determined that Hox genes essentially assign addresses to the cells of the early embryo, telling one cell it is a constituent of the front of the body and not of the rear, while informing another that it is a part of the limb destined to become a finger. The genes are particularly important for setting the body pattern from the hindbrain on down; other developmental genes are more critical for constructing the forebrain and other parts of the central nervous system.

The genes operate with great speed and efficiency, performing their remarkable patterning task in three days while the embryo is implanted in the womb, from one to two weeks after conception in humans. One reason they are able to help set the whole design up so speedily is that the embryo turns out to be constructed as a series of repetitious segments, as a new section of the body grows out as a mimic of the part preceding it, though with crucial variations that lend the final body its complexity of parts.

"This idea was initially surprising to us, because the body doesn't really look segmented, with the exception of the bones of the vertebrae," said Dr. Mario R. Capecchi of the University of Utah School of Medicine in Salt Lake City, who studies Hox genes in mice. "But from an evolutionary perspective it makes sense to design an embryo by adding on modules and then subtly changing what the module is. That way, you don't have to make up a whole new repertory for every organ, but merely to modify what already exists."

And after years in which fruit fly geneticists, blessed with an organism that is highly amenable to exquisitely refined manipulations, managed to dominate all discussions of developmental questions, vertebrate biologists at last have caught up in their understanding of what Hox genes do when sculpturing a creature with a backbone—work of obvious interest to the understanding of human development.

"At our meeting in Switzerland last summer, the International Hox Gene meeting that we hold every two or three years, I'd say the most impressive thing to come out was that the vertebrate stuff finally held its own with fruit flies," said Dr. Clifford J. Tabin of Massachusetts General Hospital in Boston, who studies how Hox genes affect the development of the limb.

Other scientists have been intrigued by the structure of the Hox genes, the startlingly precise manner in which they are organized on the chromosomes. The mystery of how the different genes labor during the crucial days of embryogenesis is just beginning to be explained, but researchers think the molecules collaborate with one another, either sequentially flicking on as development proceeds, or in some cases working in teams to assure the proper patterning of a particular part of the embryonic body.

Such genetic collusion is hardly unusual, and many tasks of life require the simultaneous or consecutive contributions of multiple genes; for example, four different genes are needed to build a single protein cage of hemoglobin, which ferries oxygen in the body.

But while the various hemoglobin participants are scattered hither and yon around the chromosomes and simply flip on as needed, no architectural organization required, the 38 Hox genes in mammalian cells are grouped in four tight clusters on four different chromosomes. In each set, the genes are lined up one after another facing in exactly the same direction and with the same spacing between them, as though a designer had drawn the plan with a T square and ruler.

Nowhere else among the 100,000 genes that make up human DNA have scientists found such an exacting configuration, and some biologists believe that this singular arrangement is no accident.

In perhaps the most brazen proposal, and one for which he admits he has scant evidence, Dr. Denis Duboule of the European Molecular Biology Laboratory in Heidelberg, Germany, suggests that the three-dimensional arrangement of the Hox genes on the chromosomes acts as a sort of built-in clock, a way of ticking off the moments as development unfolds by using the length and components of the double helix itself as the timing mechanism.

By this theory, one Hox gene turns on, produces a transcription factor that in turn kindles many other genes that together create one segment of the body; then the next Hox gene down is aroused, and it helps build a new segment of the body, and so forth, all in perfect lockstep timing. Dr. Duboule argues that in a sense the segments of the body are encapsulated in the arrangements of the Hox genes on the chromosomes, an idea with alluring artistic and philosophical undertones.

"You're using DNA to measure time, and if you think about this, it's absolutely extraordinary," he said. "It means that you have on the chromosomes a physical and temporal representation of ourselves, a representation of the body axis, of the child itself, and that's very weird. It brings us back to the idea of the homunculus."

The original homunculus theory, popular among Aristotelian philosophers, assumed that every sperm cell held a tiny human within it, which needed only the nourishment of the woman's womb to grow. And many depictions of the Annunciation, when the archangel Gabriel tells the Virgin

Mary she will bear the son of God, show a homunculus of Jesus headed toward Mary on shafts of light.

In this case, the homunculus is embedded in the four chromosomal groups of Hox genes hidden within the shadowy nucleus of the early embryonic cells.

Dr. Duboule has published his theory in *Nature* and elsewhere, and he and his students are seeking a mechanism to explain how the chromosomes just might measure time. "If you ask me how this happens, I must confess I have absolutely no idea," he said. "Most scientists tend to reject a concept when there is no mechanistic basis to explain it."

Other developmental biologists are concerned less with the arrangement of Hox genes than with how the genes perform their magic. The molecules won their name because of observations geneticists made some years ago of what happens when the genes are mutated in fruit flies.

Bombarding their insects with blistering X rays and then seeing what happened to the flies' progeny, researchers saw that some newborns emerged with bizarre anomalies, like double pairs of wings, or long legs at the top of the head where their short antennas should be. Researchers soon realized that the genetic flaws had essentially reprogrammed developmental genes, causing normal body pieces to grow where they did not belong, and they dubbed the mutations homeotic, meaning similar.

Scrutinizing these developmental genes in detail, biologists noticed that nested in each of them was an identical molecular sequence, which they designated the homeobox. The sequence is brief, a mere 183 base pairs, or building blocks, out of the thousands of DNA bases that make up individual genes. But the fact that the same motif was seen so often indicated the sequence must be a critical genetic player, and thus developmental genes that enfolded the homeobox were named Hox, after their core element.

Scientists have since learned that the homeobox is the so-called binding domain of the gene. When a Hox gene is translated into a protein to perform its chore within the cell, the homeobox specifies a little corkscrew-shape unit that allows the entire Hox protein to behave as a transcription factor, twisting around the double helix in a sinuous hug and then switching on genes.

Studies of fruit flies, however essential to the story of the Hox genes, can only go so far in explaining the making of a mammal, and vertebrate biologists have recently ventured much farther.

In the most ambitious effort, Dr. Capecchi of Utah and Dr. Pierre Chambon of the National Institute of Medical Research in Strasbourg, France, are systematically knocking out the Hox genes in experimental mice and seeing how the creatures' development is affected in the absence of one or the other of the 38 players.

They are using advanced and extremely painstaking techniques to delete the genes from embryonic mouse cells, implant the altered cells in mother mice, and then breed the resulting offspring with one another until they have so-called pure homozygous mice, rodents lacking the Hox genes.

So far, the scientists have managed to strip away 2 of the 38 Hox genes in rodents. When one of those Hox genes, the Hox A-3 gene, is deleted from embryonic cells, the resulting pup is born with a broad suite of flaws: heart defects, face and skull deformities, and the lack of a thymus, the organ where immune cells mature. Although the deformities seem dispersed and unrelated, Dr. Capecchi suggests that all the afflicted organs originated in one segment of the early embryo, and that segment was devastated by the absence of the Hox gene in charge of ensuring its initial design.

Many mysteries remain, however. Among the major questions are which genes are activated by the Hox A-3 gene, and exactly what happens when the target genes respond? Do they then turn on other genes? Do they change conditions in the embryonic cells, perhaps influencing calcium concentrations—always a theatric bit of biochemistry—or increasing the capacity of the flowering cells to respond to signals from their neighbors? Only by mapping out the entire cascade of events will scientists truly understand how the Hox genes work, and how one executive molecule like Hox A-3 can shape and knead a sliver of the embryo into the throbbing heart of a newborn child.

—Natalie Angier, February 1993

Family of Errant Genes Is Found to Be Related to Variety of Skeletal Ills

FORGET THE GREAT HUMAN HEART AND BRAIN, the eyes that see sweeping vistas, the lips that declare "I am"; the real thing standing between us and primordial ooze is the human skeleton. Built of 206 bones, 29 in the skull alone, the skeleton is a living cathedral of ivory vaults, ribs and buttresses—a structure at once light and strong, flexible and firm.

Yet like those toy kits requiring self-assembly, the skeleton is also so complicated that many things can and do go wrong with it during development. Disorders that affect the growth of the body's bones are among the most common of all birth defects. For example, at least one in 3,000 babies suffers from any of a hundred different syndromes that cause the sutures of the skull plates to fuse prematurely.

Such conditions result in heads that are too tall or too flat or grossly asymmetric or shaped like a cloverleaf, all requiring surgery soon after birth to relieve pressure within the cranial cavity. Dwarfism, syndactyly, limb deformities, the curvature of the spine called scoliosis—the list of deviations from the standard skeletal diagram is long and motley.

However, scientists have reported a spectacular series of discoveries showing that defects in the same group of genes may be responsible for many different disorders of the skeleton. They have linked disorders as varied as achondroplasia—the most common genetic form of dwarfism—and three diseases that affect the face, skull, hands and feet to mutations in a class of genes called fibroblast growth factor receptors. These genes, which were originally identified for their role in the growth of connective skin cells, or fibroblasts, have been known and studied for years. But only since 1994 have they been shown in their mutant form to be the cause of human skeletal disorders.

150

The discoveries have come so fast that even those in the field have barely been able to keep abreast of the results. The first link was announced in late July 1994, when scientists from the University of California at Irvine said that achondroplastic dwarfism resulted from a defect in the gene that allows the body to produce fibroblast growth factor receptor 3 (so numbered because it was the third of the growth factor family to have been discovered, in the 1980s).

In September, researchers from the Institute of Child Health in London announced in the journal *Nature Genetics* that Crouzon syndrome, a cause of premature fusion of the skull plates, was associated with defects in the gene that makes fibroblast growth factor receptor 2. And in another issue of *Nature Genetics,* two different teams announced yet more evidence of malfeasance among the growth factor receptor clan: In one case, scientists said a mutation in the gene for fibroblast growth factor receptor 1 causes Pfeiffer syndrome, which also leads to the untimely fusion of the cranial sutures, but in addition makes the thumbs and toes very wide. In the second, a team again linked receptor 2 to a skeletal problem, this time to Jackson-Weiss syndrome, which prompts the development of skull and foot anomalies.

"This has been the most exciting time of my entire career," said Dr. Maximilian Muenke of the University of Pennsylvania School of Medicine in Philadelphia, the lead author on the report about Pfeiffer syndrome. "From seeing my first Pfeiffer patient eight years ago, to having the gene defect known and to learning that three other skeletal defects result from mutations in the same gene family—it's absolutely thrilling."

The troika of implicated genes all do similar things in the body: they produce receptor proteins that sit on the surface of cells, including the cells of the bones, and respond to stimulatory signals from growth factors floating around them. Of great astonishment to scientists, the receptors are found on many different cell types of the body and are known to be essential to the growth and maturation of many tissues and organs; yet the impact of a mutation on the receptors' function appears to hit bone development the hardest and most consistently.

Through analyzing the genes and tracing their role in these different cranioskeletal conditions, scientists may get a handle on how independent yet related genes work together to construct the great osteotic Lego set of the body.

The findings also pose a significant challenge to the fledgling field of bioethics, however. With the mutations now identified, researchers may soon have prenatal tests available to identify an embryo with a skeletal defect early enough to permit the mother to have an abortion if she chooses. But unlike other genetic disorders that may be screened for prenatally, the skeletal syndromes in most cases are neither lethal nor excessively debilitating. They do not impair intelligence or lead to great physical pain. Instead, they make people look unusual or deformed or funny.

Hence, the findings may force scientists, ethicists and parents to confront the question of what exactly distinguishes a genetic disease from a genetic condition, an esthetic deviation from the norm, and to decide how far they are willing to go to fix what is not unequivocally broken.

"This work has caused me to think about ethical and philosophical issues more carefully than I ever have before," said Dr. John J. Wasmuth of the department of biological chemistry at the University of California at Irvine, the senior author on the report of the dwarfism gene.

The story of the receptor family began more than a decade ago, when scientists discovered a protein in the blood that prompted connective skin cells, or fibroblasts, to grow in the rarefied condition of a laboratory petri dish; given its effects, the protein was named fibroblast growth factor. Soon, another eight such factors were identified, though they all exerted different effects on different types of cells, sometimes causing the cells to divide, in other cases to mature.

In the late 1980s, Dr. Lewis T. Williams, a professor of medicine at the University of California at San Francisco and president of Chiron Technologies at the Chiron Corporation in Emeryville, California, and his colleagues isolated the protein that allows cells to respond to the growth factors' messages—the receptor for fibroblast growth factor. They and others eventually identified three more receptors, resulting in a large family of kith and kin—nine different factors, and four receptors to clasp on to the factors. Working with mice, chicks and other animals, scientists also realized that the four receptors were cut and pasted together in varying ways during embryonic development, a vivid clue that the receptors were helping to orchestrate the growth of the animal from its humble beginnings as a fertilized egg.

Dr. Williams and his coworkers learned that the different receptors turned on at different times and in different parts of the skeleton as the fetal animal grew. "This complex expression pattern indicated to us that the receptors were obviously playing an important role in bone development." Yet there were equally provocative signs that the receptors were essential to the growth of the embryonic brain and central nervous system as well.

Indeed, the signs of a neuronal role were so suggestive that a couple of years ago, when Dr. Wasmuth was seeking the gene responsible for Huntington's disease, a fatal neurodegenerative disorder, he looked at the gene that makes fibroblast growth factor receptor 3 as a good candidate. That hunch proved incorrect, but in a roundabout way it eventually led him to identify the receptor as the cause of a very different disease, achondroplastic dwarfism. This form of dwarfism is the one most familiar to people; many of the dwarf actors seen in movies and on television have achondroplasia.

The disorder is remarkably specific and affects only certain classes of bones. Dwarfs have normal-size trunks, but their foreheads are slightly enlarged and the bones of their upper arms and thighs are stunted. Most dwarfs do not inherit the condition from a dwarf parent, but instead have a so-called sporadic form of the disease—the mutation arose spontaneously in the sperm or egg cell of a normal-size parent. However, that dwarf can then beget other dwarfs, with each offspring having a 50 percent chance of inheriting the dwarfism trait.

Examining the DNA of 16 achondroplastic dwarfs from different families, Dr. Wasmuth and his coworkers were astonished to find that 15 had the same mutation in their receptor gene. And since the initial report, researchers looking at the genes of 300 dwarfs have found the identical mutation in more than 290. By contrast, hundreds of different mutations have been shown to cause cystic fibrosis, and the newly discovered breast cancer gene is likewise subject to multiple mutations up and down its lengthy span.

The consistency of the dwarfism mutation suggests to researchers that this kind of mutation causes the growth factor receptor to assume a very particular aberrant form, one that interferes with bone growth only in a limited number of bone types.

Once the association between a skeletal disorder and a fibroblast growth factor receptor had been announced, scientists studying other bone diseases leapt on the possibility that their diseases might have something to do with faulty receptors.

Dr. Muenke, for example, decided after seeing his first patient with Pfeiffer syndrome in the mid-1980s that he would someday seek the root cause of it. And when a preliminary chromosomal analysis of the disease led him and his colleagues to discover that the Pfeiffer trait seemed to be sitting exceptionally close on the chromosome to the gene for fibroblast growth factor receptor 1, they zeroed in on the receptor as the likeliest culprit for the disorder.

Sure enough, in five unrelated Pfeiffer families, the same spot of the receptor gene had been mutated, the specific genetic alteration presumably causing the distinctive symptoms of the condition. Pfeiffer syndrome is utterly different from achondroplasia, yet its characteristics are equally distinctive. In the disorder, some of the sutures of the skull fuse prenatally, rather than staying open until well after birth as they do in normal babies. (The open sutures of a baby's skull can be felt in the soft spot familiar to all parents.) As a result of the early fusion, the skull ends up growing tall and narrow, flattened in back. The forehead is large, the left side of the face may be smaller than the right, the bridge of the nose is often depressed, and the bony orbital sockets that hold the eyes are shallow, causing the eyes to look protuberant.

"You'd recognize Pfeiffer syndrome immediately once you've seen it," Dr. Muenke said. In addition, the thumb and big toe may be twice as broad as normal, and the big toes may be practically at a 90-degree angle from the rest of the foot, a deviation that may require extensive foot surgery.

The disorders resulting from defects in fibroblast growth factor 2 also give rise to abnormal skulls and, in the case of Jackson-Weiss syndrome, to broad big toes that also point toward one another like a couple of thumbs.

Even as they continue with their research, scientists are grappling with what the findings mean for their patients. The discovery of the genes puts pressure on all involved to get their ethical thoughts in order, to decide what is meant by healthy, what diseased, what normal, what unacceptable. It may be very well and good to admit that this culture places an enormous premium on attractiveness, tallness, a strong jaw and sculptured cheekbones; it is quite another to say such things are all that count.

—NATALIE ANGIER, November 1994

With New Fly,
Science Outdoes Hollywood

COMBINING ELEMENTS of the sublime and the macabre, scientists have created flies that grow large, perfectly formed eyes on the most inappropriate parts of their bodies: on their wings, on their legs, on the quivering tips of their antennae.

The experiment offers graphic evidence that scientists may have discovered what they call "the master control gene" for the formation of the eye, one of the most complex structures in nature. Researchers in developmental biology have been struggling, with scant success, to identify the genetic signals that initiate the growth of the body's specialized components, whether limb, liver or brain.

But the work, reported in the journal *Science,* suggests that the gene with which the scientists prompted laboratory flies to sprout as many as 14 eyes apiece is indeed the kingpin of vision, the gene that touches off an intricate biochemical event able to transform a nondescript speck of cells into a fully outfitted eye. Whether these extracurricular fly eyes can see remains to be determined.

Though the work was done with fruit flies, which are genetically amenable to such manipulations, the eye gene in the fly turns out to be similar to a gene identified in mammals, including humans, indicating that the equivalent gene in human embryos may direct the creation of the paired windows to the soul.

The gene used in the fly experiments is called eyeless, because the absence of the gene results in flies with no eyes at all.

"It's an amazing example of how a single gene can switch on an entire developmental program," said Dr. Walter J. Gehring of the University of

Basel in Switzerland, the senior author of the report. "It came as a total surprise to us."

Dr. Gehring estimates that at least 2,500 different genes participate in the construction of the eye, and that all those genes answer directly or indirectly to eyeless.

Dr. Gehring did the experiments with his colleagues Dr. Georg Halder and Dr. Patrick Callaerts.

"It's the paper of the year," said Dr. Charles S. Zuker, a neuroscientist and fruit fly biologist at the University of California School of Medicine in San Diego. "This is Frankensteinian science at its best."

Other scientists expressed enthusiasm for the work, though some scorned the term "master control gene," which they said was a glib phrase that ignored the highly interactive nature of the body's development, the chattering talk and crosstalk that occurs while the multitudes of growing cells figure out which does what.

"This is quite a spectacular result, but I have problems with the idea of 'master regulators,'" said Dr. S. Larry Zipursky, a fruit fly researcher at the University of California School of Medicine in Los Angeles. "I think it's an attractive way to get attention."

The new work also suggests that conventional ideas about the evolution of the eye may be wrong. In view of the vast differences between the visual systems of many different organisms, scientists had long assumed that the eye might have been invented as many as 40 different times. A human eye, with its single lens, looks nothing like the fly's compound eye, which is made up of 800 tiny eyes linked like soap bubbles in a bath.

But the paper suggests that, given the similarities between the gene for a fly eye and that for a mammalian eye, the primordial eye may in fact have evolved only once, taking on manifold shapes and designs depending on the needs of the organism. Even the squid appears to have its own version of the eyeless gene.

The fact that vertebrates like people and invertebrates like insects and squid seem to share the same master control gene for eyes "is contrary to all the textbooks," Dr. Gehring said. "I'll freely admit that that includes my own."

The latest edition of the *Encyclopaedia Britannica,* for example, discusses as a well-known fact the autonomous evolution of the squid eye and the vertebrate eye.

Dr. Hermann Steller, a geneticist at the Massachusetts Institute of Technology, said, "This convincingly draws similarities between eyes that were thought to have developed independently, and it suggests that the first simple visual system must be very old," dating back to before insects and vertebrates went their separate ways half a billion years ago, and perhaps long before that.

The first primitive eye theoretically did little more than detect differences between light and darkness, as an earthworm is able to do with the light-sensing "eyespots" found across the surface of its body. But that simple capacity presumably is all that natural selection needed to begin sculpturing the dozens of different visual systems seen in the animal kingdom today.

To build their monsterish flies, Dr. Gehring and his colleagues took copies of the eyeless gene and inserted them into regions of the primordial fly larva normally destined to become wings, legs, antennae or other body parts. When the flies hatched from their eggs, they displayed fully formed eyes wherever the eyeless gene had been installed.

Some had eyes bulging up from the middle of their wings, others on the thorax, the insect equivalent of the chest. The cutest flies were those with eyes on the tops of their antennae. "They look like little crabs, which have their eyes on stalks," Dr. Gehring said.

Nor were these misplaced eyes pale mimics of real eyes. They had all the 800 eye units. "They have the right red pigment, the little bristles between each unit eye, the lens, the light-sensitive cells inside," said Dr. Zuker, who has seen the mutants. "We're talking the works."

The Gehring team has demonstrated that the light-sensing cells, or photoreceptors, in these far-flung eyes were operative, but they must still determine if the eyes are wired to the fly's brain—that is, if the wing eyes lend the insect a new view on the world. Dr. Gehring suspects that if any eyes are likely to be connected properly, they are the ones on the antennae.

In the past, scientists have gotten animals like amphibians to grow eyes on various parts of their body by transplanting the entire embryonic optic cup from one spot to another. But the new experiments, by using a single gene to accomplish the same outcome, are far more sophisticated and revealing of the genetics of eye development.

At the moment, the work has no obvious clinical value, but scientists believe that by understanding the growth of the eye, they can eventually devise new therapies for the many visual problems that afflict people at all stages of life.

Scientists in Edinburgh are now working with mice, trying to see if they can use the fly eyeless gene to correct a genetic disorder known to leave a mouse with defective eyes. If they succeed in bestowing on the mice a healthy pair of rodent eyes, the experiment will demonstrate that the eyeless gene is a true master switch that turns on a different developmental network for eye growth, depending on the animal in which it operates.

But if, against scientific prediction, the mice somehow emerge with bristly, bloodred compound fly eyes—well, Hollywood thought of it first.

—NATALIE ANGIER, March 1995

More Than One Way, or Even Two, for a Rat to Smell a Rat

THERE MAY OR MAY NOT BE a third eye, but biologists have discovered a third system of smell in rats and mice, raising the possibility of a similar system in humans.

The new work may also help resolve whether humans possess pheromones, the hormone-like scents that arouse sexual behavior in others. Perfumes with names like Raw, Realm, Contact 18 and Desire 22 are already being marketed as containing pheromones that will make the wearer draw the opposite sex like moths to a lightbulb. But these imaginative claims are a step or two ahead of what most scientists think can be proved. Pheromones definitely govern the lives of mammals like rats and mice; their role in human behavior is an intriguing but open question.

Smell, the most primal and evocative of the senses, has a power that suggests it is wired into deeper levels of the brain than the sight and sound that dominate the conscious mind. A major new approach to understanding smell was opened in 1991 when two researchers at Columbia University, Dr. Linda B. Buck and Dr. Richard Axel, discovered a large family of genes responsible for the odor receptors in the lining of the nose. The receptors poke out of the membrane of nerve cells in the lining, known as the olfactory epithelium.

The nose has another sensory structure, called the vomeronasal organ, which is specialized for detecting pheromones. Dr. Axel and another colleague, Dr. Catherine Dulac, were surprised to find in 1995 that the vomeronasal organ in rats possesses its own set of odor receptors, specified by a different family of genes.

The story has taken another twist with the discovery that the vomeronasal organ possesses a second set of odor receptors, again with its own

159

family of genes. The finding was made independently in rats by Dr. Buck, who is now at the Harvard Medical School, and in mice by Dr. Dulac, also at Harvard University, and was reported in two articles in the journal *Cell*.

"What is interesting is that we now have three chemosensory systems which seem to be quite independent," Dr. Dulac said.

Why should nature have gone to the bother of designing three systems? Because pheromones have powerful effects, like stimulating an urge to mate as well as longer-term hormonal changes, it would seem sensible to keep the pheromone detection system separate from the sense of smell. "You don't want accidentally to induce an innate response by exposure to a pear or banana," Dr. Buck noted.

But no one knows why the vomeronasal organ should have two detection systems. They could serve pheromones with different roles, like those involved in reproduction and in social status. The only clue so far is that at least one of the second class of receptors discovered by Dr. Dulac has a different pattern of distribution in male and in female rats. The receptors also reach their final pattern only in mature animals, as might be expected with a pheromone governing sexual behavior.

In both the mouse and rat, many of the new receptor genes produce receptor proteins that seem to be dysfunctional. Some, for example, lack the section that allows the molecules to embed themselves in the cell's membrane. Dr. Buck suggests that this could reflect an evolutionary mechanism to help create new species. If animals should change their genetic pattern of pheromone receptor production by inactivating some kinds of receptors or reviving others, that would redefine whom they would mate with, and a separately breeding population could easily become a new species.

Dr. Buck and Dr. Dulac are interested in the possibility of human pheromones, but neither is convinced of their existence. There is a vomeronasal organ in humans, two little pits on either side of the partition that divides the nostrils, but anatomists have no real evidence that it is wired to the brain or sends out any signals.

"When I am in my optimistic days, I think humans are likely to have pheromone receptors," Dr. Dulac said. "But humans are not as dependent on olfactory cues, and cultural constraints are going to make any pheromonal effects much more subtle." It could be that human pheromones act together with visual signals to influence behavior. But in Dr. Buck's view, "there is no definite study that shows a pheromone in humans."

Discovery of the two families of pheromone receptor genes in the mouse and rat gives biologists a powerful new tool for tracking down whatever human pheromones may exist. Mice and men share so much of their genetic heritage that a gene in one species usually has a close counterpart in the other. Dr. Dulac has already found several human genes that are related to the new receptor genes from rats. Unfortunately, like many of the rat's pheromone receptors, the human genes seem likely, based on their DNA structure, to produce dysfunctional receptors.

Detection of the genes for the three families of receptors has enabled biologists to start figuring out how the brain identifies odors and pheromones. In the olfactory family, there are as many as 1,000 genes, each making a different kind of receptor. Each nerve cell in the olfactory epithelium seems to switch on just one of these genes, Dr. Buck has found, choosing the gene apparently at random. But a cell that has turned on a particular gene connects to a specifically positioned neuron in the olfactory bulbs, the part of the brain where smell is first analyzed.

Dr. Buck believes from her experiments that there is a system of "combinatorial coding." An odor may be detected by several different classes of receptor, each recognizing a different part of the odor's molecular structure. The cells expressing each class of receptor report to a specific set of neurons in the olfactory bulb. From the pattern of neurons that light up in the bulb, the brain can figure out what smell the nose is smelling, Dr. Buck suggests.

In the mouse's vomeronasal organ, there are 100 genes serving one family of receptors and 140 genes in the other. The numbers are similar in the rat. It is not known how many pheromones the two systems can detect.

The vomeronasal cells report to a distinct region of the olfactory bulb known as the accessory olfactory bulb. The two parts of the bulb are wired differently into the brain. Nerve fibers from the main bulb go to the olfactory cortex and then to higher centers of the brain where smells are perceived. But the accessory olfactory bulb bypasses the higher centers, sending its fibers to the amygdala, an area of emotional control, and from there to the hypothalamus, a control region for the body's hormones.

Dr. Dulac plans to explore how the nerve cells of the olfactory systems wire themselves to the brain as the animal develops. Dr. Buck's lab is exploring the human perception of smell by studying people who cannot detect particular odors. About 12 percent of people cannot smell musk, at least in low concentrations, and the same is true with mint. By examining the detection region of the receptors that detect musk, it may be possible to understand why some people cannot smell it.

"It's really not clear what is going on yet," Dr. Buck said.

—NICHOLAS WADE, August 1997

Mating Game of Fruit Fly Is Traced to a Single Gene

WILL SHE? WON'T SHE? DOES HE CARE? WILL HE CALL? Courtship is an intricate behavior in almost every species from humans to fruit flies. It energizes mind and body in life's great game of contributing to the next generation. But biologists have now discovered to their amazement that almost the entire courtship repertoire of the male fruit fly is governed by a single gene.

The gene is able to shape behavior because it gets to be switched on in the fruit fly's brain cells. Although the gene is active in only 500 or so of the 10,000 cells in the fly's brain, these few cells seem to stage-manage all the scenes in the male's courtship ritual: recognizing a female, approaching and nuzzling her, singing by vibrating his wings and consummating their union if his song is approved.

The finding raises the possibility that a related gene may operate in humans since the two species, despite their evolutionary distance, have equivalent versions of many important genes, like those that guide the development of the embryo's body plan. But even if such a gene exists, it would be unlikely to shape human behavior as strongly as the fly gene molds that of the male fruit fly.

The researchers say their discovery is apparently the first time that a gene has been found to govern a complex pattern of behavior by influencing brain cells. They hope their finding may offer a new handle on the intriguing issue of how the brain is wired up to produce a given pattern of behavior.

"In terms of the developmental genetics of behavior this would be a major step; there is not a parallel finding elsewhere," said Dr. Sean B. Carroll, a biologist at the University of Wisconsin. "I would say that's pretty startling."

Dr. Michael McKeown, an expert on fruit fly genetics at the Salk Institute in San Diego, described the research as "a very important step in allowing us to dissect what it takes to generate complex behaviors."

Fruit flies have long been a standard organism for genetic studies, and their genes have proved to have more overlap with those of higher organisms than might be expected.

The finding is the work of teams at four universities led by Dr. Bruce S. Baker of Stanford University, Dr. Steven A. Wasserman of the University of Texas Southwestern Medical Center, Dr. Jeffrey C. Hall of Brandeis University and Dr. Barbara J. Taylor of Oregon State University. A leading role in the research was taken by Dr. Lisa C. Ryner, a colleague of Dr. Baker. The scientists described their findings in the journal *Cell*.

The gene they studied has been known for many years, but its Eros-like powers have not hitherto been recognized. It was designated fru, short for fruitless, because in a strain of fruit flies with an unusual version of the gene the males lost their capacity to recognize females as the appropriate object of attention and would court one another.

Researchers at Stanford stumbled across the gene again during a 15-year program in which Dr. Baker has been studying a hierarchy of genes that determine sexual development in fruit flies. One gene makes products that act on the next gene until at the end of the chain a gene known as doublesex shapes all the embryo fly's sex organs, creating male or female equipment depending on the instructions received from genes higher in the chain.

From various hints the biologists guessed that there might be a branch in the chain, and they picked up signs of a new gene that turned out to be fru. Just as the doublesex gene shapes male and female organs, fru molds sexual behavior, at least in the male.

When defects are introduced at various points on fru, the male fruit flies lose different parts of their sexual repertoire. Further experiments showed that the gene was switched on in specific areas of the brain, putting it in a position to shape the fly's behavior.

The genetic programming of the male's courtship, and its dependence on a single gene, might make the poor fruit fly seem like little more than a robot. In fact, the gene sets more of a framework than a rigid program, since the fly has some flexibility in its routine.

The male must respond appropriately to the female's initiatives, since she may first reject him, then induce him to sing, then allow him to copulate. The two of them are carrying out genetically choreographed behavior, but they have to integrate their actions.

"These genes do establish a neural framework necessary for carrying out courtship behaviors," Dr. McKeown said, "but that behavior is engineered to allow flexibility in response to environment."

Biologists expect to find a genetic framework for human reproductive behavior, but it presumably is much looser than the system that manages the repertoire of the male fruit fly. Fru determines a male's sexual orientation, since, as noted, males with a mutated fru gene will court other males and fail to recognize females.

"In terms of what influence genes have, there is already evidence of genetic influence on human orientation," said Dr. S. M. Breedlove, a psychologist at the University of California at Berkeley, but he noted that the genetic factor was far from conclusive.

Could the genetic basis of that influence be a human counterpart of the fru gene? The researchers say in their report that mammals seem to use a different genetic system from that of fruit flies for determining an organism's sexual development. But biologists have found that at least one of the genes in the fruit fly hierarchy, a gene called tra-2 that helps control fru, has evolutionary counterparts in both mice and humans. The role of the gene in these species is not yet understood.

Dr. Baker said that no counterpart of the fru gene had yet been found in the rapidly growing data banks of human gene sequences, but that could be because the gene was switched on so rarely that it might not have been caught by the gene sequencers.

"Given this finding in flies, I wouldn't be surprised to find people use the same genetic logic for their sexual behavior," he said. "We know you build the body plan with the same genes, and evolution may have taken care to see that reproduction was well done and well coordinated by being under control of a single gene."

Besides its relevance to courtship, discovery of the fru gene's properties is likely to intrigue biologists who study the wiring pattern of the brain, since it points to the existence of a system for wiring a pattern of behavior into an organism's brain.

The essence of the fru gene system is that it produces several different protein messages, depending on the instructions it gets from tra-2 and another gene above it in the hierarchy. Biologists call the trick alternative splicing because the usual relationship is for each gene to specify a single protein product.

Fru's protein product seems to be what is called a transcription factor, one that switches on a target set of genes. The researchers have not yet discovered what the effect of fru is in the female fruit fly, perhaps because the female's courtship behavior is more elaborate and less well studied than the male's.

—NICHOLAS WADE, December 1996

Modern "Wolfmen"
May Have Inherited Ancient Gene

CASTING A SLENDER RAY OF LIGHT on the mysteries of both hair growth and the legend of the werewolf, scientists have discovered a gene that in its mutant form causes hair to sprout thickly and thoroughly across the face and upper body, covering the cheeks, forehead, nose and even the eyelids.

The rare hereditary condition, called congenital generalized hypertrichosis, results in such a furry appearance from birth that scientists propose it could be an example of an atavistic mutation—the reemergence of an evolutionarily ancient trait that is normally kept suppressed. In this case, the mutation harks back to the prototypically mammalian state of near-total hirsutism, the possession of a protective fur coat that modern humans for some reason lost at an unknown point in the past.

In fact, those with generalized hypertrichosis—*hyper* meaning "excess," and *trichosis,* "hair"—are even hairier than chimpanzees or gorillas, which lack fur around the cheeks, nose and eyes. The only parts of the patients' bodies without hair are the palms of the hands and the soles of the feet. This suggests that the atavism could recall something earlier than the emergence of hominoid apes about 25 million years ago.

"This is probably a mutation of a gene that was a sleeping beauty," said Dr. José M. Cantu, head of genetics at the Mexican Institute for Social Security in Guadalajara, an author of the report. "The mutation awakened a gene that had been put aside during evolution."

But Dr. Cantu and his colleagues emphasized that the idea of generalized hypertrichosis as an atavistic mutation was only a theory. "At this point it's strictly speculation, though the idea is a very interesting one," said Dr. Pragna I. Patel of the Human Genome Center at Baylor College of Medicine in Houston, another author of the report, which appeared in the journal *Nature Genetics.*

Biologists have observed many other mutations that they suggest fall into this class of atavisms, the reappearance of normally dormant traits. Some people are born with multiple sets of nipples, for example, just as most nonprimate mammals have a double ridge of mammary tissue down the length of the underside of the torso. In very rare cases, girls develop entire extra breasts at puberty.

Other examples of atavistic mutations include the extension of the human coccyx into a small tail, the appearance of hind limbs in whales and the growth of extra toes on horses and cats.

"Atavistic mutations tell us that a lot of information is kept around for a very long time," said Dr. Brian K. Hall, a developmental biologist at Dalhousie University in Halifax, Nova Scotia. "Just because an animal isn't using a gene anymore doesn't mean the information just disappears." Dr. Hall wrote a commentary about atavistic mutations that appeared with the report on hypertrichosis.

The researchers do not yet have the precise gene isolated, but merely know its approximate location, on the bottom half of the X chromosome. They found the location by examining the genetic material of a large Mexican family, whose members may be the only humans known to have this particular mutation.

In the past, scientists have described other types of hypertrichosis. However, the newly detected mutation has a few outstanding features. For one thing, those who have the condition show no abnormality other than hirsutism. In many other types of hereditary hypertrichosis, there are other disorders beyond too much hair, including facial and skeletal abnormalities and mental retardation.

Moreover, the Mexicans with this form of hypertrichosis have exceedingly thick and abundant body hair. In other conditions, the hair is often finer or less complete in its coverage. Finally, because this newly described hypertrichosis is a so-called X-linked trait, it finds its fullest expression in males. Females with the condition inherit only one copy of the mutant X chromosome, and so their other, normal copy of the X chromosome gives them partial protection against the disorder. As a result, they display only patchy spots of excess hair growth.

By contrast, if a boy inherits from his mother the X chromosome with the hypertrichosis mutation, he has no second X to give him some buffer against its effect. With the mutation active in every cell of his body, he ends up with a more uniform coat of hair.

The researchers said that, apart from any insights it might give into the evolution of humans, the identification of the gene could offer clues to how hair grows, of which scientists currently have very few. Presumably, the same or similar molecular signals involved in generalized hypertrichosis also control the more cosmetically accepted form of hair growth, on the scalp. Once the gene proper is isolated, and the protein it makes is understood, the work could in theory lead to a better treatment for baldness.

Homing in on the hypertrichosis gene may not be easy, given the extraordinary rareness of the condition. What is more, Luis E. Figuera, a doctoral student working in Dr. Patel's laboratory who collected the blood samples from the family members to perform the genetic research, had a difficult time persuading many people with the trait to participate in the study. Some thought the possibilities of a treatment for the disorder were too remote to bother taking part, while others were accustomed to keeping their distance from the public. In the end, Mr. Figuera persuaded about half the 32 family members to donate blood for genetic screening.

Some of the boys and men in the family work mightily to get rid of their excess hair, shaving every part of their face every day, sometimes repeatedly. But a couple of them have decided to use their spectacular appearance to earn a living and have joined circus sideshows in Mexico.

People with this and other types of hypertrichosis have been stigmatized throughout history, and they are thought to be the source of the werewolf legend. They have often been displayed in circuses as "dog men," "ape women," "human Skye terriers" or "*Homo silvestris,*" man from the woods. The most famous of the hairy "human curiosities" was Julia Pastrana, a Mexican Indian born in the 1830s who had long, thick, glossy and straight black hair covering most of her body, as well as overdevelopment of the jaw and an unusually broad, flattened nose. Scientists have tentatively diagnosed Pastrana's condition as "generalized hypertrichosis terminalis with gingival hyperplasia," a different condition from that discussed in the *Nature Genetics* report, though with the same extraordinary degree of hairiness.

Pastrana was exhibited to huge throngs throughout the United States and Europe by her husband, Theodore Lent, a tireless if unscrupulous impresario. Lent continued to profit from his wife even after her death— first by selling her corpse to a Russian anatomist, then by getting back her beautifully mummified body from the anatomist and exhibiting it. The mummy was long believed lost, but in 1990 it was rediscovered at the Oslo Forensic Institute in Norway, where it remains.

The identification of atavistic mutations also dates to the 19th century. Darwin suggested that polydactyly—having more than the usual number of digits—could reflect an ancient trait intermittently reappearing as a result of some hereditary or developmental misstep. The horse of Julius Caesar was believed to be one such atavar, for contemporary documents describe its hoof as cloven in five parts and looking like a human foot. Biologists have also observed that some whales are born with vestigial hind limbs, indicating that the instructions for making legs remain even though whales, when they returned to the sea 40 million years ago, lost their need for limbs.

Apart from mutations reactivating dormant genes, atavisms appear often during embryonic development and then disappear before birth, said Dr. Hall. For example, the duck-billed platypus, among the oddest of mammals, grows teeth in utero as most mammals do; but at some stage in its embryonic development the teeth disappear and are replaced by a horny beak.

Biologists propose that the reason atavisms exist at all is nature's propensity for recycling old ideas. Rarely is a gene used for a single purpose in the growth and health of an organism. Instead, most genes are Renaissance artists, able to work in a range of styles and media depending on the needs of the species. A gene involved in hair growth may also play a role in the development of skin or bones. Thus, even a relatively naked ape like *Homo sapiens* cannot afford to lose the hair gene for fear of jeopardizing the rest of the body's architecture and packaging.

—NATALIE ANGIER, May 1995

6

GENES AND DISEASE

The promise of genetics has always been that it will furnish a fundamental understanding of genetic diseases such as cancer, and that from the understanding will develop better treatments and even cures.

The premise is true in principle, but it has been a long time coming to fruition. Biologists have gained remarkably deep insights into the genetic mechanisms that go awry in cancer cells. But gene therapy, the concept of introducing normal genes to correct a genetic defect, has proved extremely hard to put into practice.

Thus the discovery of genes like BRCA1 and BRCA2, which in defective form cause the hereditary forms of breast cancer, is a great advance in knowledge but is of little practical help until methods are available to replace or counteract the defective gene.

On the other hand discovery of a gene that predisposes carriers to colon cancer could save many lives since through annual examinations of the colon, precancerous growths, known as polyps, can be detected and removed.

In the case of genetic diseases caused by several genes, it is much harder for the causative genes to be isolated by means of family pedigrees. Thus diseases like schizophrenia are known to have a strong genetic component, but the underlying genes remain unknown.

Advances in genetics also help explain the fundamental nature of infectious diseases, in particular the interaction between invading microbes and their host's immune defense system. Viruses and mammals are adversaries in an ancient war and have evolved stratagems of amazing subtlety against each other. Viruses have learned how to burrow into human cells and lie low, waiting their time to attack. The immune system has developed means of detecting and killing the cells that harbor the invaders. All these stratagems are programmed into the genes of human cells and of the microbes that prey on them.

Breast Cancer Gene
in One Percent of U.S. Jews

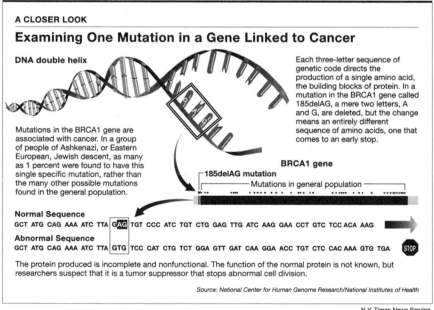

A CLOSER LOOK

Examining One Mutation in a Gene Linked to Cancer

DNA double helix

Mutations in the BRCA1 gene are associated with cancer. In a group of people of Ashkenazi, or Eastern European, Jewish descent, as many as 1 percent were found to have this single specific mutation, rather than the many other possible mutations found in the general population.

Each three-letter sequence of genetic code directs the production of a single amino acid, the building blocks of protein. In a mutation in the BRCA1 gene called 185delAG, a mere two letters, A and G, are deleted, but the change means an entirely different sequence of amino acids, one that comes to an early stop.

BRCA1 gene

185delAG mutation

Mutations in general population

Normal Sequence

GCT ATG CAG AAA ATC TTA GAG TGT CCC ATC TGT CTG GAG TTG ATC AAG GAA CCT GTC TCC ACA AAG

Abnormal Sequence

GCT ATG CAG AAA ATC TTA GTG TCC CAT CTG TCT GGA GTT GAT CAA GGA ACC TGT CTC CAC AAA GTG TGA

STOP

The protein produced is incomplete and nonfunctional. The function of the normal protein is not known, but researchers suspect that it is a tumor suppressor that stops abnormal cell division.

Source: National Center for Human Genome Research/National Institutes of Health

N.Y. Times News Service

IN A NEW INSIGHT into the genetic basis of breast cancer, biologists have discovered that a particular genetic defect is found with unusual frequency in American Jews whose ancestors lived in Eastern and Central Europe.

The discovery does not seem likely to lead immediately to any new treatment, but it could result in a better understanding of the incidence of breast cancer and might explain the apparent slight excess of breast cancer that some studies have found among Jewish women.

It also raises the difficult question of whether it makes sense to screen healthy people for the defect, given that there is now no good therapy to offer those in whom it is found.

The defect is a mutation, or change in DNA units, of a gene known as BRCA1, which was found a year ago to be associated with breast cancer cases that run in families. Familial breast cancer accounts for 5 percent to 10 percent of all breast cancer cases. The gene seems to play a pivotal role in suppressing malignant changes, and when it is inactivated, the breast cells become much more prone to cancer.

At first, researchers hoped to develop a simple test for the mutation that inactivated the BRCA1 gene. They soon found that many different mutations could occur at various sites along the gene, each inactivating the gene in different ways.

The new discovery concerns a special feature of a particular mutation, known as 185delAG. As many as 1 percent of the descendants of Ashkenazi Jews, those of Eastern and Central European origin, have this mutation in their BRCA1 gene. In effect, this amounts to about 1 percent of American Jews, since 90 percent to 95 percent of them are of Ashkenazi descent. This is a surprisingly high percentage for a genetic disease, since even quite common genetic diseases are usually found in a very small proportion of the population.

"One percent—it is a shock to think that that could be true," said Dr. Judy Garber, director of cancer risk and prevention at the Dana-Farber Cancer Institute in Boston.

"I'm totally stunned," said Dr. Francis S. Collins, director of the National Center for Human Genome Research in Bethesda, Maryland, who is an author of the study. "This becomes, at least for the Jewish population, the most common genetic disorder. It is hard to think of anything close to it as far as a single gene that causes an illness."

The study is the work of a team led by Dr. Lawrence C. Brody of the National Cancer Institute and was published in the journal *Nature Genetics*.

The health significance of the new finding is at present uncertain and the subject of an urgent new study. Women with a family history of breast cancer and a mutated BRCA1 gene are at considerably increased risk: The mutation confers an 80 percent to 90 percent chance of getting breast cancer and a 40 percent to 50 percent chance of developing ovarian cancer. There are also hints that it might lead to colon cancer and, in men, prostate cancer.

But these risks are not necessarily true for people who have the mutation but no family history of the cancer. It is possible, for example, that other genes may counteract the mutation's effects in people with no family history of the disease. Researchers have started a survey, which will take at least six months, to examine this question.

The new discovery raises an ethical dilemma that geneticists have long discussed: whether to test people for the mutation. Suddenly, Dr. Collins said, geneticists are confronted with issues they thought they could wrestle with over a period of years. If the gene proves to be a powerful predictor of cancer risk, should all Jewish women be offered a test to see if they have it? Who is going to counsel them, and how? How will they be protected against discrimination by employers and insurance companies?

Genetic testing has recently begun for people with family histories of cancer. But no one has offered the tests to people who had no reason to suspect that they were at risk. Now the possibility arises of offering a screening test to millions of healthy people. There are six million Ashkenazi Jews in the United States.

Unlike tests for other mutations, which often involve laborious searches of huge genes for evidence of minute changes, this test is easy because it involves looking for a single alteration of a gene at a known position.

"Boy, this changes everything," Dr. Collins said. "We thought we would have another three or four years to wrestle with issues of ethics and the delivery of genetic services to the general population. All of a sudden, those years just got taken away."

Dr. Michael M. Kaback, a geneticist at the University of California at San Diego, said: "This is an enormously important intersection on the road. This will be a testing we've never had before."

Previous studies had suggested that the rate of breast cancer among Jewish women might be 10 percent to 20 percent higher than the rate among non-Jewish women, but some researchers had suggested that the increase, if real, might be due to diet or other nongenetic factors.

The discovery of the BRCA1 gene in 1994 was hailed as a first step toward understanding the molecular events that can cause cancers of the breast and ovaries. But investigators cautioned it was not likely to lead to genetic screening.

The problem was that the gene was only known to be associated with the 5 percent to 10 percent of breast cancers that are familial and it seemed to have an almost unending variety of possible mutations. The only way researchers could know whether a mutation was dangerous was to know that in that woman's family, others who had the same mutation had developed cancer.

But several investigators soon stumbled upon an unexpected finding: Several of the Jewish families who had participated in the research leading to the discovery of BRCA1 had exactly the same mutation. It resulted from a small deletion of genetic material that causes the production of a truncated protein, destroying the gene's function.

"What was intriguing was that the other mutations were all over the place, but the Ashkenazis only had this one," Dr. Kaback said.

Dr. Mary-Claire King, a geneticist at the University of Washington in Seattle, said that so far, every Jewish family that had a BRCA1 mutation had this mutation. She said she knew of 18 families that had it.

Dr. Brody soon proposed a study to see how prevalent the mutation was in the entire Jewish population. There was a ready sample of cells from people who had been screened to see if they were carriers of genes for other inherited diseases.

As many as 35,000 Jews are tested each year to see if they carry the mutation for Tay-Sachs disease, a deadly neurological disorder. Thousands more are tested for the cystic fibrosis gene. Non-Jews are tested for hemophilia and Duchenne muscular dystrophy if those diseases run in their families.

In these diseases, people with a single copy of the mutated gene are healthy. But if both parents carry the gene, their children would have one chance in four of inheriting a copy from each parent and so getting the disease. Many couples choose to abort affected fetuses. In contrast, a single copy of a mutated BRCA1 gene confers a cancer risk.

Dr. Brody and his colleagues obtained cells from Dr. Kaback, whose center tests about 15,000 Jews each year for Tay-Sachs genes, and from two other testing centers. They removed all identifiers so that no one could know which individuals turned out to have the BRCA1 mutation. For comparison, they tested cells from people who had the hemophilia or muscular dystrophy gene tests.

That led to their finding that 1 percent of Ashkenazi Jews appear to carry the mutation. They calculated that they would expect about 1 in 800 non-Jews to have the mutation, a result that is not contradicted by their failure to find it in the single sample of 815 non-Jews.

Now, the researchers said, come some hard questions: How dangerous is the mutation in people without a family history of breast cancer? And what can people carrying the mutation do to protect themselves from dying of cancer?

It is possible that the mutation is not as potent in the general population as it is in families with a strong history of cancer. But, Dr. King said, the gene "is a classic BRCA1 mutation" and researchers have good evidence that a nonfunctioning BRCA1 gene is important in the development of breast cancer.

To sort this out, the cancer institute is immediately beginning a study of 4,000 to 5,000 Jewish men and women in the Baltimore and Washington areas, testing them for the BRCA1 mutation and collecting extensive family histories to see if cancers are more frequent in the families of those who have the mutation. Dr. King and Dr. Joan Marks of Sarah Lawrence College in Bronxville are conducting a similar study in New York.

Cancer specialists said that even if the mutation turned out to be a powerful predictor of cancer risk in the general population and even if Jewish women were offered a screening test, it was still not clear what women who had the mutation gene could do to protect themselves.

Dr. Barbara Weber, who directs the cancer risk evaluation program at the University of Pennsylvania, said that she offered screening at six-month intervals for breast and ovarian cancer to women from families with high rates of cancer who had mutated BRCA1 genes. Some of these women want their breasts and ovaries removed, a drastic step that still does not confer complete protection from the cancers, Dr. Weber said. But, she added, if, after counseling, they insist on the operations, her center will comply.

Because no one knows yet how to prevent cancer in women with a BRCA1 mutation and no one knows yet whether early detection makes a difference in these women, some, like Dr. King, are opposed to screening tests. "We need better options," Dr. King said. "We need to get out of the purgatory of identifying people and then not knowing what to do about it," she said.

—GINA KOLATA, September 1995

Gene Mutation Tied to Colon Cancers in Ashkenazi Jews

IN A DISCOVERY that offers a new way of preventing many cases of colon cancer, biologists have detected a genetic change that apparently doubles the risk of the disease.

The genetic change, or mutation, occurs in as many as 6 percent of people of Ashkenazi Jewish descent, according to preliminary studies, making it the most common known cancer gene in a particular population. The mutation has not yet been found among non-Ashkenazis.

Biologists at the Johns Hopkins Oncology Center in Baltimore, where the mutation was found, have also developed a test to detect it. The center is recommending that anyone of Ashkenazi descent with a close relative who has had colon cancer should take the test, because people with a family history of the disease are at higher risk of developing it themselves.

The Lerner Foundation of Cleveland is offering to pay for the test for anyone of Ashkenazi descent who cannot afford it. But the test is not yet available for people living in New York State, since the state department of health has not approved any laboratory to offer it.

Ashkenazi Jews are those who settled in Central and Eastern Europe. Most Jewish Americans are of Ashkenazi descent.

The discovery has two novel aspects. It suggests, if its initial premises are confirmed, that there may be substantial medical benefits in testing a whole population group, even though this would raise risks of discrimination in jobs and insurance, and even of stigmatization. Second, the mutation appears to cause cancer by a completely new mechanism, a finding that may lead to the discovery of other cancer-causing mutations that work in a similar way.

Dr. Francis S. Collins, director of the Human Genome Project at the National Institutes of Health, described the discovery as "pretty dramatic." But he suggested that the mutation's significance needed to be further evaluated. "This is a new kind of medicine," he said, "and I worry about plunging into it too quickly. The public will not be amused if we overestimate the risk of this."

He pointed out that the risk associated with two breast cancer genes had been overestimated at first.

Dr. Collins also said the finding did not mean that people of Ashkenazi descent faced a greater burden of genetic disease. Some mutations are simply easier to find in certain populations because of what geneticists call a founder effect. "There is no perfect genetic specimen," Dr. Collins said. "We are all flawed, we all carry five to fifty serious genetic misspellings. That one has now been identified in Ashkenazis doesn't change the bottom line that we are all walking around with things that are probably a lot worse than this."

People who are found to carry the new mutation could have regular colonoscopies, a procedure in which doctors examine the lower intestine and at the same time remove any polyps, the slow-growing, finger-like protrusions from which colon cancer originates.

Colonoscopies are not cheap or pleasant, but they substantially reduce the risk of colon cancer.

"We could expect a benefit as high as ninety percent," said Dr. Sidney Winawer, meaning that at least nine out of ten potential cases of cancer could be averted. Dr. Winawer, an author of the report, is an expert on familial cancer at the Memorial Sloan-Kettering Cancer Center in Manhattan.

If the initial figures for the frequency and risk of the mutation are confirmed, some experts believe that the test should be offered to everyone of Ashkenazi Jewish descent. The general lifetime risk of colon cancer among this population group is 9 percent to 15 percent, but it is estimated to be twice that—18 percent to 30 percent—for the 6 percent of Ashkenazis who are believed to carry the mutation.

As a practical matter, however, the Johns Hopkins Oncology Center, the only place where the test is now available, can screen limited numbers. This is one reason the test is not now being recommended more broadly. The other is the need to quantify the risk better.

"We are certainly recommending that anyone with a family history of colon cancer be tested," said Dr. Bert Vogelstein, a member of the center and a chief author on the research team that discovered the mutation. "In other cases, at least initially, we are leaving it up to the patients and their physicians."

Dr. Vogelstein added, though, that "anyone who wishes to get the test, we won't refuse, but I think it is an individual decision."

Alfred Lerner, whose foundation has agreed to pay for the test, said he hoped a lot of people would take advantage of that opportunity. Asked how much the program might cost, he said: "I really don't want to put a number on it. Let's just see how it plays out."

Dr. Vogelstein believes that the discovery has the potential to save thousands of lives, because in the United States alone an estimated 360,000 people, or 6 percent of the six million Ashkenazi Jewish Americans, carry the mutation. "None of these patients need to get seriously ill if they have the knowledge," he said. "This is a totally preventable illness."

The colon cancer mutation was discovered by a team led by Dr. Vogelstein and Dr. Kenneth W. Kinzler, also of the Johns Hopkins Oncology Center, and was reported in the journal *Nature Genetics*. The team includes Dr. Steven J. Laken, also of the center, and colleagues at several other institutions. Dr. Vogelstein is widely known for having unraveled the series of genetic changes by which a colon cell becomes cancerous.

The mutation was discovered in an individual, not even a patient, who was paying a social visit to Dr. Vogelstein. The visitor, who had a family history of colon cancer, asked to be tested to see whether he had a genetic predisposition to cancer.

Dr. Vogelstein agreed as a favor to test the visitor's DNA, but found only a single, harmless-looking change in a gene known as the APC gene. That gene specifies a vital protein that suppresses tumors, and is known to be mutated in a kind of familial cancer, known as adenomatous polyposis, where numerous polyps appear in the colon. But in that case the APC gene produces a shortened and ineffective version of its protein, because the mutation introduces an erroneous stop signal in the middle of the gene's instructions to the cell's manufacturing machinery for making a protein chain.

The visitor did not have familial polyposis and the mutation in his APC gene did not produce a stop signal, but instead merely changed the code specifying an amino acid unit known as lycine to isoleucine. Since the two are similar building blocks of a protein chain, Dr. Vogelstein at first judged the mutation to be a harmless common variant known as a polymorphism.

But recalling a colon cancer patient who had had the same mutation as his visitor, Dr. Vogelstein looked farther and found it in many other colon cancer patients. Like the adenomatous polyposis mutation, the new mutation was also producing truncated APC proteins but by a novel mechanism.

The mutation, though innocuous in its direct coding change, converts a single chemical letter to one identical to neighbors on either side, generating a string of eight identical units that confuses the cell's replication apparatus. Each time the cell divides, it is liable for unknown reasons to miscopy the sequence by inserting an extra chemical letter before or after it. Occasionally, the inserted letter creates a stop signal that truncates the APC protein and sabotages its tumor suppressor role.

The creation of a mutational hot spot is a previously unknown biological mechanism that may allow many other seemingly harmless changes to be recognized as cancer-causing mutations.

"This has to be the tip of the iceberg, since nothing is ever unique in genetics," Dr. Vogelstein said.

The APC hot-spot mutation is not the only cancer-predisposing gene that is more common among people of Ashkenazi descent. The two breast cancer gene mutations are found in about 1 percent of Ashkenazi women. Several other diseases, like Tay-Sachs and cystic fibrosis, are also more common among Ashkenazis. The higher incidence is thought to be due to what is called the founder effect.

"The world population of Ashkenazis went through a real bottleneck that lasted from 800 A.D. to 1400 or 1500 A.D.," said Kenneth Offitt, a clinical geneticist at Sloan-Kettering. "During that time the population contracted well below two million and was broken into small communities with intermarriage."

The mutations that happened to be common among the founders of the community are still common and occur at higher frequencies than do mutations in larger populations among whom genes are more widely mixed.

Larger populations tend to have a greater number of founder mutations but none is very common.

"I think it is very unlikely that the total number of genetic aberrations carried around by Jewish individuals is any greater than that of any other group," Dr. Collins said. "Other groups may have a larger number of founders and a greater range of mutations, so that in the aggregate they carry just the same burden of disease."

Nonetheless, two clinical geneticists who cosigned Dr. Vogelstein's report expressed concern about how an expanded testing program might be perceived both outside and within the Ashkenazi community.

—NICHOLAS WADE, August 1997

Doctors Isolate a Common Cancer-Related Gene

AN IMPORTANT GENE that predisposes people to various kinds of cancer as well as causing a rare genetic disease has been isolated after an intensive search lasting more than a decade.

The discovery, reported in *Science,* has aroused strong interest among researchers because, unlike other cancer-related genes, the defective form of this gene is relatively common in the population, suggesting it may be a significant cause of many cancers.

The gene is also the cause of a dreaded genetic disease. Children who inherit the defective form of the gene from both parents suffer from a disease of movement that strikes at an early age and is usually fatal by age 20. Why the defect should cause this disease as well as cancer is still unknown.

Identification of the gene will not lead immediately to any therapy but it affords a deep insight into the nature of cancer, one from which practical ways to reverse the defective gene's damage may eventually be developed.

As many as 1 percent of Americans, or more than two million people, carry the defective form of the gene, which is thought to increase their risk of getting any of a variety of cancers three- to eightfold. The cancers associated with the gene, called the ATM (for ataxia telangiectasia mutated), include those of the breast, lung, stomach, skin and pancreas.

In contrast, previously discovered cancer genes, including those that predispose to colon cancer or breast cancer, affect just one in several hundred to one in one thousand people.

"It's extremely important," said Dr. Bert Vogelstein, a molecular biologist at the Johns Hopkins University School of Medicine. Dr. Vogelstein said that, for breast cancer alone, there are twice as many patients with the ATM gene as with the BRCA1 gene, the recently discovered cancer gene that received a torrent of publicity.

Dr. David Housman, a molecular biologist at the Massachusetts Institute of Technology, said that compared to the retinoblastoma gene, which causes tumors in 90 percent of those who carry it, the ATM gene was relatively ineffective in causing cancer. But, he said, what makes the ATM gene so important is the number of people who have it. Although it increases cancer risk by just severalfold, the numbers soon add up, making it a major cause of cancer.

For cancer researchers, who have been looking for ways to find particularly susceptible groups to test new methods of preventing cancer, the ATM gene could be an important new lead. Dr. John Minna, director of the cancer center at the University of Texas Southwestern Medical Center at Dallas, said the discovery of the gene was "fantastic" and could give a huge boost to his studies and those of others who are testing ways to prevent lung cancer.

But the discovery also gives rise to questions of how the information will be used, and by whom. Once again, while basic research leaps ahead, society is faced with the dilemma of whether to tell people they have the gene and, if so, what to tell them. While scientists hope to go slowly, using the gene to investigate the molecular biology of cancer and to select participants for studies on new ways to prevent cancer, companies are already anticipating selling a genetic screening test that would inform people of their cancer risk.

A troublesome problem raised by the ATM gene is that although it increases a woman's chances of developing breast cancer, it also may raise her risk of having cancer induced by radiation in mammograms. But there is no reason yet to believe that this hypothetical risk outweighs the demonstrated benefits of mammograms in large populations, said Dr. Francis S. Collins, who is director of the National Center for Human Genome Research in Bethesda, Maryland, and an author of the paper on the ATM gene.

The finding also can benefit a much smaller group of people, those who inherit two copies of the ATM gene, one from each parent, and so develop the terrible disease known as ataxia telangiectasia. Its symptoms include neurological problems, skin lesions, a weakened immune system and a predisposition to cancer. Most patients with the disease die by age 20.

Dr. Collins said that although the disease, occurring in about 1 of 40,000 Americans, is just half as common as Huntington's disease, it is "a horrible disease" and families "haven't had much hope until now."

The disease has intrigued researchers for years, with its puzzling symptoms, coupled with a connection to cancer. Some, including Dr. Michael Swift of New York Medical College in Valhalla, spent years studying the parents and first-degree relatives of children with ataxia telangiectasia, asking whether the gene also predisposes them to develop cancer. He reasoned that each parent must have a gene for the disease and that the relatives also were likely to have a gene for the disease. These studies were the basis of the estimates that a single copy of the ATM gene increases cancer risk severalfold.

For children with two copies of the gene, the disease was inevitable and they faced a "a steady, downhill, tragic course," Dr. Collins said. One of the hallmarks of ataxia telangiectasia is that cells of patients are extremely sensitive to radiation. Their skin cells, tested in the laboratory, are unable to repair themselves in the way that normal cells do.

The reason for this sensitivity is now clear, said Dr. Yosef Shiloh, a geneticist at Tel Aviv University in Israel, who is the lead author of the paper. Usually, a cell that sustains damage to its DNA stops in its tracks and repairs itself. But cells with the ATM gene ignore the damage and continue to replicate their DNA and divide. This means that some of these cells would be so damaged that they would soon die, whereas others could become cancerous.

Dr. Vogelstein said that this supports a growing body of evidence indicating that the most important cancer-causing genes seem to be those that inhibit a cell's ability to repair its DNA.

Dr. Collins said he expects that the first application of the isolation of the ATM gene will be to diagnose the disease in children, anticipating that a test will be available "in a year or two." Eventually, Dr. Collins said, the gene might lead researchers to treatments that might slow or ameliorate the disease.

Brad Margus of Boca Raton, Florida, who has two children with ataxia telangiectasia, said that many children with the disease are undiagnosed or misdiagnosed. His six-year-old son, Jarrett, first showed signs of the disease when he was two, stumbling and developing slurred speech. Mr. Margus and his wife, Vicki, went from doctor to doctor trying to learn what was wrong. Only when Jarrett's younger brother, Quinn, started showing the same symptoms did doctors suspect that the boys might have an inherited disease.

Mr. Margus, who is president of the A-T Children's Project, a foundation that promotes research on the disease, said that as the isolation of the ATM gene became imminent, he began hearing from companies that wanted to market it as a screening test for cancer susceptibility.

—GINA KOLATA, June 1995

Big Picture of Cancer Process
Is Being Seen for the First Time

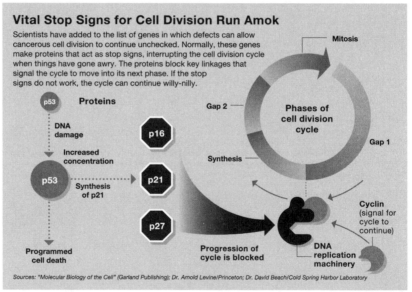

Vital Stop Signs for Cell Division Run Amok

Scientists have added to the list of genes in which defects can allow cancerous cell division to continue unchecked. Normally, these genes make proteins that act as stop signs, interrupting the cell division cycle when things have gone awry. The proteins block key linkages that signal the cycle to move into its next phase. If the stop signs do not work, the cycle can continue willy-nilly.

p53   **Proteins**

DNA damage

Increased concentration

p53   **Synthesis of p21**

p16

p21

p27

Programmed cell death

Gap 2

Phases of cell division cycle

Gap 1

Synthesis

Mitosis

Cyclin (signal for cycle to continue)

Progression of cycle is blocked

DNA replication machinery

Sources: "Molecular Biology of the Cell" (Garland Publishing); Dr. Arnold Levine/Princeton; Dr. David Beach/Cold Spring Harbor Laboratory

N.Y. Times News Service

FOR YEARS, molecular biologists have held fast to a deep-seated belief about the nature of cancer. Malignancy, they argued, must have something to do with a loss of the molecular brakes that halt the growth and division of cells. After all, they reasoned, the most distinctive feature of a cancer cell is that it does not stop growing, while a normal cell does.

If so, then any defects in the genes that control the cell cycle—the cell's orderly progression from growth, to division into two daughter cells, and back to the resting state—would be a likely cause of cancer. This grand hypothesis sought to unify two major research programs, study of the cell cycle and the search to understand the molecular basis of cancer.

Yet it was harder to prove than anyone had expected. Much of the evidence "wasn't really compelling," said Dr. Bert Vogelstein, a molecular geneticist at Johns Hopkins University. "When people actually measured cell cycle parameters in cancer cells in the body," Dr. Vogelstein said, "they did not find any defects. It seemed reasonable that there was some problem with the cell cycle, but exactly what wasn't clear."

But that picture has recently changed. Basic research on the cell cycle and studies of the genetic causes of cancer have now built to "a crescendo of results that are really bringing the cell cycle right to the heart of the cancer problem," Dr. Vogelstein said.

The crescendo reached a peak with a paper in *Science* magazine announcing a newly discovered cancer gene that turns out to be a major gene in the cell cycle. In its cell-cycle role, which was first discovered in 1993, the gene makes a protein product called p16, which helps cells to stop growing.

Two important cancer genes discovered in the last year by Dr. Vogelstein and Dr. Richard Kolodner of the Dana-Farber Cancer Institute in Boston derange the cell cycle only indirectly. These genes produce an enzyme that repairs errors in DNA; when the genes are defective, mutations accumulate throughout the cell, including possibly in the cell-cycle genes.

Another important cancer gene, p53, acts indirectly on the cell cycle by preventing cells from making proteins that they need to stop growing.

The finding has caused a buzz of excitement among biologists because the gene in question specifies a protein that governs the cell cycle. Its defective form has been identified in a wide range of human cancers.

The finding, molecular biologists say, together with the rich lode of research over the last several years, has elevated scientists' understanding of cancer to a new level.

"What has happened is that we're beginning to understand for the first time the big picture," said Dr. Arnold Levine, a molecular biologist at Princeton University. By elucidating how cell division is regulated and, more important, how it is stopped in normal cells, and by seeing how these molecular events are disrupted in cancer, the pieces of the cancer puzzle are snapping into place, he explained.

"You have to understand before you can design ways to intervene," Dr. Levine added.

The new p16 discovery stands out because it is such a strong confirmation of the theory and because the gene seems to be involved in such a wide variety of cancers that it is quite likely to be a central figure in the drama of a cell's spiraling changes toward malignancy.

"The real importance of this work is that it allows us for the first time to ask questions about designing drugs to target" the specific molecular defects, Dr. Levine said.

The work relies on the newly detailed understanding of how the cell cycle is controlled. In the orderly rhythm of the cycle, first cells pause, then they copy their DNA and pause again, and then they divide. Researchers named these phases of the cell cycle G1, S, G2, and M, with *S* standing for "synthesis" of DNA, *M* for "mitosis," or cell division, and *G1* and *G2* standing for "gap" 1 and 2. Dr. Levine said gap was meant to convey "a time gap and also a gap in our knowledge."

In studying the gaps over the past few years, researchers found, at each phase of the cell cycle a class of proteins, called cyclins, starts to accumulate until they reach a critical mass. At that point the cyclins alter another cell protein that in turn pushes the cell to the next phase of its cycle. One cyclin buildup follows the next, driving the cell to move from phase to phase.

The next question was, how does the cell shut this process down when its division has been accomplished? Researchers knew that if a cell runs out of nutrients, or if its DNA is damaged, its growth cycle comes to a sharp halt. But how is that halt brought about?

In the case of damage to DNA, researchers now know that a protein called p53 is the critical agent in commanding that growth and division stop. The cell makes p53 all the time, but the protein is usually broken down almost at once. But when the cell's DNA is damaged, whether by chemicals or X rays or sunlight, the usual breakdown process ceases and p53 builds up.

An accumulation of p53 is a critical signal that forces a stark choice on the cell. If the DNA damage is substantial, the cell essentially gives up on the idea of repairing it and commits suicide. P53 turns on the synthesis of proteins that commit the cell to self-destruction. When the damage is less severe, p53 switches on the synthesis of another protein that in turn blocks the cyclins so the cell cycle is frozen in place. The cell then has time to repair its damaged DNA before it starts to divide.

The protein that blocks the cycle of cyclin accumulation was discovered recently and named p21. Given its central role in controlling the cell cycle, it is little wonder that a defect in the gene that specifies it can lead to the cycle's spinning out of control.

But DNA damage is not the only reason cells stop dividing. Hormones can stop them; so can starvation and some viral infections. Yet no matter what the cause of the halt to cell division, the type of molecular steps the cell takes to put on its brakes is the same.

For example, a hormone called transforming growth factor beta can stop cell division. It does so by signaling the cell to make a protein, called p27, that binds to the cyclins and stops the cell cycle cold, blocking the progression from the G1 to the S phase.

When viruses infect cells, they can only multiply when the cells are growing and dividing. So one protective mechanism that cells devised is to block a cyclin and stop dividing. In fact, Dr. Levine said, it was by probing a tumor virus infection in hamster cells that he and his colleagues discovered p53. He found that the hamster cells began making p53 to stop dividing and so protect themselves from the virus. But the virus had developed a counterstratagem of blocking the cell's p53, effectively destroying the cell's brakes on its division. The result was that the virus caused cancer.

Mutations in p53 have now turned up in about half of all human cancers, indicating that with the inactivation of this gene human cells, too, can progress toward malignancy.

The path to the p16 discovery was opened last year by Dr. David Beach of Cold Spring Harbor Laboratory on Long Island. While studying the controls of the cell cycle, he found that a protein, which he called p16, always seemed to be clinging to the cyclins. Its role turned out to be to block cyclins so that the cell cycle is halted. It was the same effect caused by p53 through the mediation of the p21 protein, as well as by transforming growth factor beta through another blocking protein, p27.

Dr. Mark Skolnick and Dr. Alexander Kamb of the University of Utah and Myriad Genetics in Salt Lake City came upon the cancerous version of the p16 gene by an entirely different route—by looking for the gene that causes melanoma. At least 10 percent of melanoma cases arise in families that inherit a predisposing gene. As they closed in on that gene, Dr. Skolnick and his colleagues discovered that it was a mutated version of the just discovered p16 gene. They also found that the gene could become mutated in people born with a normal form of it.

In retrospect, Dr. Levine said, "you could have predicted it." A gene like p16 that directs the synthesis of such an important molecular brake almost had to be a cancer gene if the cell-cycle theory of cancer was correct. But the fact that Dr. Skolnick and his colleagues stumbled upon the gene "was sheer coincidence," Dr. Levine added.

Discovery of the role of the p16 gene in cancer has focused attention on the genes for the two other cyclin-blocking proteins already identified, p21 and p27. An obvious conjecture is that defects in these genes can also give rise to cancer.

"People are going crazy now looking for mutations" in the p21 and p27 genes, Dr. Beach said. "An absolute floodgate has been opened."

—GINA KOLATA, April 1994

Can the Common Cold Cure Cancer?

ON A SUNNY San Francisco morning in the University of California's gleaming new Cancer Center, Frank McCormick is listening to the complaints of David Jablons, a thoracic surgeon. Jablons is describing the lack of good remedies for lung cancer with a candor that few patients get to hear.

"The majority of what we do is depressingly pathetic—for the most part cancer is a systemic disease," Jablons says, referring to its frequent dissemination through the body even before the first tumor is noticed. "For advanced-stage disease, lung cancer is 90 to 95 percent fatal when diagnosed."

McCormick is director of the Cancer Center, where he is responsible for coordinating the work of 250 researchers. Jablons wants to treat some of his own patients with a novel therapy devised by McCormick—one of a new generation of cancer treatments based on a dawning understanding of how cancer cells actually work. Strangely enough, McCormick's approach closely resembles something already found in nature; his experimental cancer therapy is a cunning adaptation of a virus that causes the common cold.

Early results, from patients with cancers of the head and neck, look promising: In the toughest cases, patients for whom the standard forms of treatment have failed are already seeing their tumors shrink and occasionally disappear. "This is potentially very exciting stuff," Jablons says, as he sits at a small table in the director's office. The surgeon is eager to try out McCormick's therapy on lung-cancer patients, but he must first test it to get approval from his hospital review committee.

In 1971, Richard Nixon declared the first all-out war on cancer. The goal was to conquer the disease with a blitzkrieg program similar to the *Apollo* moon shot, and the crusade was based on the premise that human cancers were caused by a virus and could be eliminated by a vaccine made from the killed virus. So sure of success were the planners that they built a high-security laboratory on the Bethesda, Maryland, campus of the National

Institutes of Health to study the human cancer virus just as soon as it was isolated.

The cancer war, though overly optimistic, was still not a complete waste of money. Believing that cancer had to be understood before it could be conquered, biologists used some of the funds for a project that at the time had no obvious clinical relevance: the study of viruses that caused tumors in animals. Little by little, over many years, researchers teased out an extraordinary scientific story. These viruses, it emerged, were able to make an infected cell plunge into the uncontrolled growth typical of cancer because they had stolen copies of some of the cell's own genes, subtly corrupted them and then undermined the cell's own restraints on uncontrolled growth.

Human cells are so complex that it might have taken biologists decades to pinpoint just which subset of the cell's 70,000 genetic instructions gets garbled in cancer. Instead, it turned out that a handful of weird viruses had done the job for them, because the genes that the viruses subvert in animal cells are close counterparts of the ones that go awry in human cancer.

Some human cancers are caused by viruses, but most begin in other ways. Genetic instructions get garbled through inheritance, or in the miscopying of DNA when a cell divides or from physical damage wreaked by carcinogens like tobacco smoke, ultraviolet light and radiation. The errors are usually harmless, except when they occur in the genes that control cell growth.

Now, more than a quarter century after the war on cancer was declared, biologists have at last accumulated much of the basic information needed to wage such a war effectively. That the cancer doctor's three main tools—surgery, radiation and chemotherapy—are often of so little use is no surprise: A disease caused by genetic instability requires a genetic remedy.

But a fourth wave of cancer therapy is about to break. Frank McCormick has long been at the forefront of developing the new understanding of how a cell's genetic controls go awry in cancer, and in 1992 he founded a company, named Onyx Pharmaceuticals, to put the new knowledge to practical use. While at Onyx, McCormick came up with a remarkable scheme for redirecting adenovirus, a cause of the common cold, to strike at cancer cells instead of the normal cells that line the nose and throat. This, and similar treatments being devised by other scientists, may or may not succeed, but at least they are finally attacking cancer at its genetic origins.

On weekends, Frank McCormick likes to risk his neck racing a Formula Mazda at a track near his home in San Francisco. The skill in racing cars, he explains, lies in taking the corners at maximum speed. "Your mind tells you you can't make it, you are going too fast, but you have to keep your foot down on the accelerator. You need absolute concentration, lap after lap." He has won several races for drivers over age 40, a class considered disadvantaged because of its greater fear. After a race, he says, "I feel mellow for several days."

Competition looms large in his day job, too. Cancer research is a world in which university scientists and biotech companies hotly pursue a limited number of leads—and the road is littered with failed approaches and unfulfilled promises.

Though McCormick heads cancer research at one of the world's leading universities, he doesn't fit the usual mold of biomedical grandee. He doesn't have an M.D. degree, for one thing, and although well known in academic research, he has spent most of his career in the private sphere. He prefers T-shirts, and his mop of prematurely gray hair is often in need of a comb after he has been zipping around San Francisco in his Mercedes convertible.

McCormick was born in 1950 in Cambridge, England, where his mother's father was a professor of ancient Egyptian history, but he grew up in Duckhole, a village in Gloucestershire. Hoping to see something of the world beyond the confines of rural England, he joined Voluntary Service Overseas, the British version of the Peace Corps, and was assigned to the town of Navrongo in the remote northern savannas of Ghana.

There he taught math and science at a Catholic missionary school, but it was the diseases he encountered that gave him the idea of becoming a doctor. Back in England he applied to medical school but was advised to take a course in biochemistry first. That diverted him into a career in molecular biology, first at Cambridge University and then at the State University of New York at Stony Brook.

In 1981, while McCormick was doing postdoctoral research in molecular biology at Berkeley, he was offered a job at the Cetus Corporation, one of the first private biotech companies. The biotechnology industry was barely out of its infancy, and his academic friends were surprised by the move, given how beholden scientists in industry were to the dictates of the market. How was McCormick ever going to build up his scientific reputation under those circumstances?

But McCormick flourished at Cetus. He managed to maintain high visibility by discovering a component of an important cell-signaling system known as the Ras pathway. The pathway's ordinary role is to accept messages from outside the cell indicating that the body needs the cell to divide into two. But when genes in the Ras pathway become mutated, the signaling system can get jammed in the on position. Under continual instructions to divide, the cell becomes a tumor.

McCormick became head of a group seeking drugs to block the Ras pathway, to prevent tumors from growing, and his work on the project won him worldwide attention.

But as a biotech company, Cetus was a brilliant failure. It supported high-quality work like McCormick's research in cell signaling and the invention by Kary Mullis of the PCR technique, a Nobel Prize–winning method of analyzing DNA. But in 1991, after a failed gamble on the anticancer drug interleukin-2, Cetus was bought by the Chiron Corporation, another biotech company. McCormick, by then a vice president of Cetus, decided to leave and form Onyx.

At the new venture, McCormick became chief scientific officer and set about trying to translate recent research findings about cancer cells into actual treatments. The group he brought from Cetus continued to work on the Ras pathway, while new groups were formed to search for drugs aimed at the errant genes known to cause breast and colon cancer.

In December 1992, McCormick conceived a novel idea for an anticancer agent: a crippled version of an adenovirus. That a disabled virus, cause of the mildest of human diseases, could become a weapon against the deadliest seems far-fetched—until you consider that adenoviruses and cancer face the same problem. They can't cause their respective diseases unless they subvert the checks that restrain human cells from uncontrolled division and growth.

Two of the principal checks are proteins named Rb and p53, nerdy labels that designate remarkable pieces of biological machinery. Rb, so called because it was first discovered in a rare cancer called retinoblastoma, clamps itself to the agents that drive the cycle of cell division; normal cells cannot divide unless Rb releases its grip, prompted by the cell's receipt of an appropriate growth signal.

The p53 protein, named after a measure of its size, can detect when its cell's DNA has been compromised, whether by damage or by foreign DNA like that of an infecting virus. The p53 protein will then force the cell to destroy itself for the common good.

Cells become cancerous when, over the course of a person's lifetime, carcinogens damage the genes of sentinel proteins like Rb and p53. The adenovirus has an even quicker way of tripping living cells into uncontrolled division: It is equipped with two genes, known in virology-speak as E1a and E1b; E1a makes a molecular missile that homes in on Rb, and E1b's mission is to sabotage p53. With its Rb gatekeeper disabled, the infected cell starts to divide, at which point the inactivated p53 cannot initiate the cell-suicide program that would otherwise be triggered.

But five years ago, the role of E1a in countering Rb was merely suspected. "The question that came up in my mind was, how do you prove that that is E1a's function?" McCormick recalls. It occurred to him that a rigorous proof would be to take two different cells—one that lacked a good Rb gene, like a retinoblastoma cell, and an adenovirus that lacked a working E1a gene—and show that the crippled virus could grow in an Rb-crippled cell but not in a normal cell. The idea was a clever double negative: The disarmed virus should be able to kill the disarmed cell.

Then came the lightbulb moment. The long-desired goal of all cancer therapy is an agent that can discriminate between cancerous and normal cells. "In thinking that through, I realized that such a virus would only grow in cancer cells," McCormick says. An adenovirus without its E1a gene could not harm normal cells. And a virus lacking its E1b gene would thrive in cancer cells that had knocked out their own p53, but not in healthy cells.

After thinking about the idea overnight, McCormick decided he had something. "I remember going to Onyx the next day," he says. "My colleagues said: 'That's it. Do it.'"

An idea conceived to solve an arcane problem in virology had suddenly become a potent method, at least in principle, for attacking cancer cells. But McCormick now needed an adenovirus with its E1a or E1b gene disabled, so he hired a consultant to call around to university virology laboratories and see if they had already developed such viruses for research purposes.

The focus soon shifted from E1a to E1b, the gene that attacks p53, since E1b is easier to work with. The Onyx consultant discovered that Arnold Berk at the University of California, Los Angeles, had developed an E1b-crippled adenovirus, and Berk willingly provided samples.

Now called Onyx-015, the virus was found in 1996 to rid experimental mice of tumors, justifying preliminary trials in patients. Many biologists were skeptical, saying the body's immune system would kill the virus before it could kill the tumors. "I got cold feet many times during its development," McCormick says of the Onyx-015 project, "because it seems such an odd-ball thing to be doing. I felt like I was out there without any clothes on."

Onyx went public in 1996. At the time, McCormick's stake in the company amounted to 189,242 shares (worth around $1.1 million as of December 11, 1997), but the exertions of running a growing enterprise were beginning to outweigh the rewards. "I felt responsible for everything that happened," McCormick says, "from the lightbulbs to the receptionist's behavior." Worst of all, he never seemed to have time to do new experiments, so when he got an attractive offer to reenter the world of academic research after 15 years away, he was ready to accept.

McCormick joined the University of California, San Francisco, in January 1997 as head of a division within the new Cancer Center. The university did not find a director for the whole center until May, when it decided that the best person to be McCormick's boss was McCormick himself.

One afternoon a week, he drives his Mercedes across the bay to consult for Onyx at its offices in Richmond. At a recent monthly review meeting there, David H. Kirn, the company's director of clinical research, reviewed the Onyx-015 trials with McCormick and the other scientists on the project. The Phase 1 trials, in which a drug's safety for patients is assessed, are complete. Phase 2, designed to find the correct dosage levels, has already begun. And ahead lies the decisive test: Phase 3 trials with large numbers of patients.

The mood around the table in the company's drab conference room was one of barely suppressed elation. The group is engaged in what McCormick calls "a battle between emotional hope and intellectual doubt." Most cancer drugs lead nowhere, and promising results at the early stages of clinical trials are often cruelly shattered during full-scale trials. Yet Onyx-015 has made an interesting start.

The Phase 1 patients, who were treated in Texas and Glasgow, Scotland, had large tumors of the head and neck that were not responding to standard treatments like surgery, radiation and chemotherapy. The Onyx virus was introduced directly into the tumors—one of the first times a genetically engineered replicating virus was injected into humans as part of a clinical trial—and by the end of the first phase, six out of nine patients saw their tumors shrink by as much as two thirds.

Kirn notes that the injected tumors in the Phase 1 trials had died on the inside but not around the edges. The researchers wondered if the patients were developing antibodies against the virus, just as would be expected in people fighting a cold, and the antibodies were protecting the outside rim of the tumors. Or maybe virus injected at the center of a tumor was simply not reaching the periphery.

The company then tried a new delivery technique, with the virus injected in and around the rim of the tumor as well as inside. In the Phase 2 trials that are now under way, Kirn says, the technique seems to be working. In one of the trials, the Onyx virus is being administered alone; in another it is being tested together with the two standard chemotherapeutic agents for head and neck cancer. "The new injection technique has given us the ability in principle to totally eradicate the tumor and its rim," Kirn says, although he declines to give the success rate.

Back in his office at the cancer center, McCormick discusses with Jablons the latest Onyx results. Both know that at the first stages of testing, when the company sponsoring a new drug may want it to work almost as desperately as the patients, everything is stacked in favor of success, a result that often cannot be repeated.

"Early in clinical trials it's always the first patients who are the best," McCormick says.

"Luckily Wall Street hasn't figured it out yet," Jablons jokes.

McCormick recounts how during safety tests of Onyx-015, company doctors anxiously monitored the patients' livers, which they feared the new therapy might damage. "One of the first patients came in with high liver enzymes and scared the hell out of us," he says. "But it turned out he was feeling so good after his treatment that he had been to the pub and spent an evening drinking. The whole nature of clinical research is hoping for the best. That's why it has such a bad rap among lab scientists."

But still he can't help hoping. Onyx-015 could well prove useless in the final test, or it could end up as a niche treatment, helpful only in special cases. The dream is that it could become the first cancer treatment based on a true understanding of how cancer cells actually work.

Whatever the outcome, McCormick finds it bracing to be in the middle of the race, taking the curves as fast as he can.

—NICHOLAS WADE, December 1997

In the Rush Toward Gene Therapy, Some See a High Risk of Failure

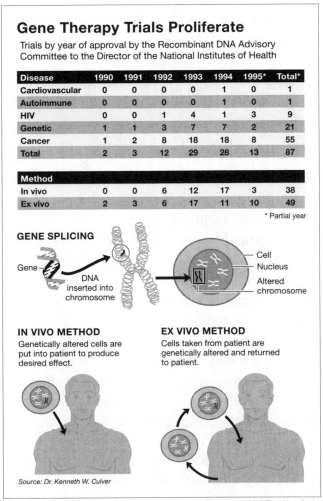

Gene Therapy Trials Proliferate

Trials by year of approval by the Recombinant DNA Advisory Committee to the Director of the National Institutes of Health

Disease	1990	1991	1992	1993	1994	1995*	Total*
Cardiovascular	0	0	0	0	1	0	1
Autoimmune	0	0	0	0	1	0	1
HIV	0	0	1	4	1	3	9
Genetic	1	1	3	7	7	2	21
Cancer	1	2	8	18	18	8	55
Total	2	3	12	29	28	13	87

Method							
In vivo	0	0	6	12	17	3	38
Ex vivo	2	3	6	17	11	10	49

* Partial year

GENE SPLICING

Gene

DNA inserted into chromosome

Cell
Nucleus
Altered chromosome

IN VIVO METHOD
Genetically altered cells are put into patient to produce desired effect.

EX VIVO METHOD
Cells taken from patient are genetically altered and returned to patient.

Source: Dr. Kenneth W. Culver

N.Y. Times News Service

IN THE SIX YEARS since the first corrective genes were injected into the first patient, gene therapy has exploded. As many as 600 patients have been given gene therapy in studies, for diseases ranging from rare metabolic

201

defects to common cancers. Nearly half of these patients were enrolled between January and June 1995.

But for all the frenzy, there has not been a single published report of a patient who was helped by gene therapy. And technical questions have multiplied. For example, scientists have discovered that the immune system may attack and destroy cells that were treated with gene therapy, seeing them as foreign or, when the genes were transferred by a virus, as infected.

The venture capitalists who have set up small companies that are a driving force behind this new field are presumably shrewd judges of the risks they are taking. Yet a number of leading scientists, including the director of the National Institutes of Health, say they worry that the field is rushing too fast from the laboratory to the bedside, driven in part by scientists' desire to climb aboard the bandwagon and in part by commercial interests that have little to do with medicine.

While others assert that it does not hurt to rush ahead with something as exciting as gene therapy, the critics say that they fear a public backlash and a crash of the gene therapy biotechnology industry if study after hastily planned study is inconclusive and if, as so often happens in medicine, some patients are injured in therapeutic disasters.

The promise of gene therapy is that by inserting corrective genes into patients' cells by a variety of ingenious methods, researchers will be able to cure inborn diseases, like cystic fibrosis, that are caused by a faulty gene. Another therapeutic technique would use genes that make a cell self-destruct. Researchers hope to infiltrate these genes into cancer cells or cells infected with the virus that causes AIDS.

Gene therapy is the most fundamental type of cure because it can correct defects in the body's own machinery. Drugs or surgery are crude by comparison. And so even the harshest critics of today's studies do not doubt that, someday, gene therapy will transform medicine. But, they say, that day will be years and maybe decades from now. In the meantime, they add, the entire enterprise is threatened by exaggerated expectations of instant cures and by clinical trials that have been designed so hastily that they are unlikely to be of value.

Scientists said there were immense pressures to rush forward with gene therapy studies. In universities, said Dr. James M. Wilson, director of the Institute for Human Gene Therapy at the University of Pennsylvania, scientists crave the prestige of being "the first in your university to do gene therapy." And, he said, universities want to announce that they are part of the gene therapy wave. "A lot of universities want to get involved," Dr. Wilson said. "It is high visibility, something people consider the future of medicine."

But the biggest force behind the gene therapy wave, some say, has been biotechnology companies. Sixty-two percent of the gene therapy studies that have begun are financed by industry.

Dr. Harold E. Varmus, director of the National Institutes of Health, has appointed an outside committee of experts, none of whom would be major beneficiaries of a change in the pattern of financing gene therapy research, to advise him on the directions the field is taking. He said that the NIH was spending $200 million a year on gene therapy research, including very basic science. Dr. Varmus wants to be sure the money is being put to good use. And he is uneasy.

"I think there's a dynamic here that's hurtful to the whole enterprise," Dr. Varmus said. "It's driven in part by company concerns," he added, explaining that companies want to have gene therapy studies so that they can show investors that they are part of the new gene therapy gold rush. Companies "gain obvious credibility in their attempts to raise money," Dr. Varmus said. "It helps in raising capital to say they actually have something in progress."

Dr. Stuart Orkin, a molecular biologist and hematologist at Children's Hospital in Boston, said that the push had been to just get clinical trials going, whether or not they were likely to have any meaningful results. Dr. Orkin, who is cochair of the panel that Dr. Varmus appointed, said that "the general impression is that there are a very large number of clinical trials, many of which have been approved without extraordinary oversight in terms of scientific or clinical benefit."

And, Dr. Orkin said, from the businesses' point of view, clinical benefit may not be so important.

"From the companies' standpoint, from a purely business standpoint, most are interested in selling out," Dr. Orkin said. "They want to be bought out or taken over. Whether they deliver a product is not germane. They just need to recoup their venture capital."

But others say that it is not so clear that the clinical trials are pushing ahead too quickly.

Dr. Malcolm Keith Brenner, a gene therapy researcher at St. Jude's Children's Research Hospital in Memphis, said that if anything, gene therapy was not moving quickly enough. "You've got to take what you've got and make the best of it," he said.

"As things improve, fine," he added, "but in the meantime, there are things that can be done."

Dr. M. James Barrett, chief executive officer of Genetic Therapy Inc. of Gaithersburg, Maryland, said that "the best is sometimes the enemy of the good in medicine."

"In our view, it is certainly appropriate to proceed with what we have," he said. In July 1995, the Swiss drug company Sandoz announced that it was buying Genetic Therapy for $295 million.

Dr. John Coffin, a molecular biologist at Tufts University and a member of Dr. Varmus's committee, took a middle ground. "There is a valid argument to be made on both sides," he said. "There is always a range of opinions as to how fast to push something." Pushing too slowly, he noted, can get researchers mired in technical minutiae, while pushing too fast can waste opportunities to perfect the method.

Dr. Coffin added that the advantage of having companies as the driving force in testing gene therapy was that "it's a mechanism by which research can be advanced fairly quickly."

So far, companies have not seemed overly worried by the risks of moving too fast. For example, Genetic Therapy was involved from the beginning in clinical trials of a gene therapy for glioblastoma, a deadly brain cancer. Dr. Kenneth W. Culver and Dr. R. Michael Blaise of the National Institutes of Health got the idea for the treatment several years ago. "As soon as the data looked good in animals, GTI wanted it," Dr. Culver said.

On December 7, 1992, Dr. Culver, Dr. Blaise and their colleagues at NIH tried the method in the first patient with glioblastoma. In the next two and a half years, 14 more patients received the treatment. "Some didn't respond, some had some response, and one man has gone two and a half years without evidence of a tumor," said Dr. Culver, who has since left the NIH and is currently between jobs.

Asked whether he thought the treatment was a success, Dr. Culver said, "My conclusion is that this therapy had no toxic effects and some evidence of biologic effects." He said that although it was highly unusual for a patient to live for two and a half years after being treated for a glioblastoma, it was not unheard of, so he could not say for sure that the treatment cured the man who had done so well.

Dr. Culver thinks it may be too soon to launch into a clinical trial. "It is my opinion as a scientist and a physician that we've got a lot more we can learn and do to optimize this type of gene delivery and destruction of tumor cells," he said. When a clinical trial begins, he noted, refinements of the method are no longer of interest. "Once the decision is made the refinement issues are only scale-up issues," he said.

But Dr. Barrett, Genetic Therapy's chief executive, said that the initial results were promising enough for the company to go ahead with a study of up to 30 patients. The study began in the fall of 1994, Dr. Barrett said, and includes patients who, unlike the original 15, have tumors that are operable.

Of course, gene therapy may turn out to be a dramatic cure for brain cancer. But more likely, said Dr. Mark Levine, a cancer specialist who is chief executive officer of the Hamilton Regional Cancer Center in Ontario, such treatments being tested now will, at best, shrink tumors slightly.

A clinical effect like that "is far from saying it's better than what we do now," Dr. Levine said. But, he added, he fears that because gene therapy "is high tech, there will be this tremendous drive to use it without having properly evaluated it."

Others worry that there is still so much to be learned about getting genes into cells that many studies are doomed. Researchers generally are not able to transfer enough genes into cells and, in some cases, the cells that get new genes will be destroyed by the immune system as foreign.

Dr. Wilson, in fact, halted a gene therapy study for precisely these reasons. He had given gene therapy to eight out of thirty cystic fibrosis patients he had planned to treat when he realized that he simply had to go back to the lab and work on gene delivery.

That was in 1993, and by 1995 Dr. Wilson said he thought he had learned enough to choose an optimal way to deliver genes within a few months.

Some of Dr. Wilson's patients were shocked by his decision to stop his study and begged him to give them the genes anyway, arguing that they had not long to live and nothing to lose. But he refused.

Dr. Wilson said he was fortunate because his work was paid for by the NIH and the Cystic Fibrosis Foundation, and not by a company. Stopping a study, he said, "is extremely expensive" and choosing a new way of delivering genes may not be feasible when patents are involved.

For experts in the field, difficulties such as this are reminders that the real issue is not when, but how. Gene therapy, Dr. Varmus said, "will happen eventually." But, he added, "the question is, do we do it right?"

—GINA KOLATA, July 1995

Artificial Human Chromosome
Is New Tool for Gene Therapy

BIOLOGISTS have accumulated wonderfully detailed insights into the genetic causes of human disease but in many cases still lack the tools to make use of the knowledge. A treatment of great promise, gene therapy, has long been stalled by the lack of suitable vehicles to get genes into human cells and maintain them in working order. But that obstacle could dissolve if a strange recent biological invention works as well as its makers hope.

The invention is an artificial human chromosome.

Synthesizing just the essential components of a human chromosome, researchers found that human cells would assemble the inserted pieces into a working chromosome, which could serve as a vehicle for therapeutic genes, they reported in the journal *Nature Genetics*.

Gene therapy holds such promise in combating inherited diseases, heart disease and cancer that the federal government has been pouring $200 million a year into research, and the pharmaceutical companies' research efforts are said to be even larger.

But so far, almost all attempts to introduce curative genes into patients have fizzled. A report prepared for the National Institutes of Health in December 1995 concluded, "While the expectations and the promise of gene therapy are great, clinical efficacy has not been definitively demonstrated at this time in any gene therapy protocol." Little has changed since, said a coauthor of the report, Dr. Arno G. Motulsky of the University of Washington in Seattle.

A major reason is that the stripped-down viruses used in many experiments to move human genes into cells do not provide a stable platform for the genes to operate. The viruses are good at sneaking into cells, as nature designed them to do, but some are not large enough to carry a full human gene and all its genetic control switches, while others provoke attack by the immune system.

So most introduced genes have lacked stable expression, the biological term for a gene producing its protein. The artificial human chromosome could sidestep many of these problems because it promises to provide a stable, natural platform for human genes to function and to be passed to daughter cells when the parent cell divides.

Commenting on the development, Dr. James M. Wilson, director of the Institute for Human Gene Therapy at the University of Pennsylvania, said, "Stability of expression is a big deal, and if we can in fact create a genetic element that functions like a chromosome, it is a really important advance."

Dr. W. French Anderson, director of the gene therapy laboratories at the University of Southern California School of Medicine, described the advance as "fascinating science" that would be "very helpful for the development of gene therapy vectors." He predicted, however, that improved viral vectors were likely to be in use before artificial chromosomes could pass manufacturing and regulatory hurdles.

The artificial chromosome is the fruit of more than a decade's work by Dr. Huntington F. Willard, a human geneticist at the Case Western Reserve University School of Medicine, together with colleagues at the university and at Athersys, a Cleveland biotechnology company.

The project is based on the insight that if the minimum essential components of a human chromosome were to be introduced into a cell, the cell's natural repair system might be able to stitch them together into a working whole.

The Cleveland team synthesized the ends of human chromosomes, short runs of DNA known as telomeres, and a long stretch of DNA designed to correspond to the centromere, a vital structure toward the center. Human centromeres, the staging posts for the apparatus that spins dividing chromosomes apart during cell division, are still somewhat mysterious. Synthetic versions have not been made until now, although several groups have been trying.

The synthetic telomeres and centromeres and a long test region of human DNA were all inserted separately into test-tube cultures of human cells with the help of a fat-based chemical that helps DNA squeeze through cell membranes.

As the researchers had hoped, not only did the cells' repair systems knit together the fragments of DNA, but the assembled DNA molecule also became clothed in chromatin, the special proteins characteristic of chromosomes.

The human cells treated the artificial chromosomes like one of their own, dutifully furnishing a copy to each daughter cell throughout the 240 cell divisions of the six-month experiment.

Dr. Willard and his colleagues, in their report, called the development an important step "toward building a prefabricated artificial chromosome capable of introducing and stably maintaining therapeutic genes in human cells." His colleagues were Drs. John J. Harrington, Gil Van Bokkelen, Robert W. Mays and Karen Gustashaw.

The Cleveland biologists believe that human genes built into the artificial chromosomes would operate normally, but they have not yet tested the idea. The DNA used in these preliminary experiments carried a marker gene, which was expressed, but no known human genes. "We have every reason to expect human genes would be expressed just fine," Dr. Willard said.

Dr. Van Bokkelen, president and chief executive of Athersys, the company that hopes to exploit the invention, said the immediate plan was to improve the artificial chromosomes as vectors for moving genes into cells. After that, the technology might be used to treat blood cell disorders like sickle-cell disease, hemophilia and immune deficiencies. These diseases have long been a target of gene therapy because blood-forming cells can be removed, treated and injected back into the body.

In the longer term, Dr. Van Bokkelen said, the new technique could be applied to developmental diseases like muscular dystrophy and cystic fibrosis and to forms of hereditary cancer.

Among the advantages of the artificial chromosomes, he noted, are their mitotic stability, which means that they are reliably reproduced and passed on every time the cell divides. There appears to be no immediate limit to the number of genes and regulatory elements that could be loaded onto them, he added.

Some cancers develop when a cell's tumor suppressor genes get knocked out of action by a mutation. The ample carrying capacity of the artificial chromosome makes it possible to think of packing it with spare copies of tumor suppressors and other disease-preventing genes, along with switches to turn them on should the natural copies fail.

But Dr. Van Bokkelen said his company was considering gene therapy only for somatic cells, the ordinary tissues of the body, and not for germline cells, the ones that give rise to eggs and sperm. "There is definitely a boundary there that the National Institutes of Health and conventional medicine has decided it isn't going to cross," Dr. Van Bokkelen said. "The idea of germline therapy is something people aren't ready even to consider at this point."

Experts in gene therapy generally agreed in interviews that the creation of artificial human chromosomes was a valuable scientific advance and held the theoretical promise described by the Cleveland team. But they cited the hurdles that remain to be cleared, like understanding the components of the artificial chromosome better and developing a more efficient way of inserting the DNA into cells.

"One of the great appeals of using artificial chromosomes," said Dr. Alan E. Smith of the Genzyme Corporation in Cambridge, Massachusetts, "is to introduce all the subtleties of the genetic controls that exist at the chromosomal level."

Artificial chromosomes were first created in yeast, a model organism much studied by biologists, in the early 1980s, but efforts to do the same for humans have long been frustrated. Building a working centromere was the missing piece.

"Before this paper, no one could do it," said Dr. Ronald G. Crystal, a gene therapy expert at New York Hospital–Cornell Medical College. "It's like when people first put genes into cells twenty years ago—it's a starting point."

Gene therapy aside, Dr. Willard regards the artificial chromosomes as a means toward understanding how chromosomes work. "One of the questions that has dogged me for the ten to fifteen years I have been in this business is that of what chromosomes are," Dr. Willard said. "We are totally ignorant of the rules that apply here. By manufacturing chromosomes, we can try to understand nature's rules."

—NICHOLAS WADE, April 1997

Gene Hunters Pursue Elusive
and Complex Traits of Mind

IT WAS JUST A HANDFUL of years ago that biologists were waving their spears, shields and pipettes in the air, boasting with full-throated glory of their success in capturing the legendary prey of molecular genetics. In a series of widely publicized discoveries, geneticists announced the isolation of the cystic fibrosis gene, the gene for Lou Gehrig's disease, the gene for Huntington's disease and a gene linked to the familial form of breast cancer.

Some of those great gene hunts had taken a decade or more, and had been a Ninth Circle of Hell for many a graduate student and postdoctoral fellow. But molecular geneticists now look back on such triumphs, shake their heads and say, boy, did those lucky devils have it easy.

The field of genetics is moving into a new and much more difficult phase: the search for genes that may contribute in some partial and numbingly convoluted way to complex traits of the mind, the stuff of private psyche and inner life. A fat set of reports published in the journal *Nature Genetics* includes two studies that cover this territory.

One confirms an earlier report that had linked male sexual orientation to a spot on the X chromosome. Another package of papers from several international teams of scientists identifies the rough location of a gene that may play a role in schizophrenia. The reports are accompanied by an editorial that tells biologists how to discriminate between a real finding in the complex field of complex traits and an experimental coincidence no more meaningful than, say, flipping five heads in a row.

This new work in genetics is riddled with scientific, intellectual and sociocultural minefields. Many of the straightforward disease genes have been isolated, or are on the verge of being so. These are the genes that hew to the tidy laws of Gregor Mendel, the father of modern genetics. These are the genes that almost surely cause illness in people born with defective versions of them. Inherit the cystic fibrosis mutation, for example, and you are at grave risk of chronic, debilitating lung infections. About 4,000 diseases are thought to be so-called single-gene disorders. Most are very rare in the population, but they have appealed to human geneticists because they are linked to one gene apiece, and are therefore open to molecular dissection.

Scientists are now moving on to the dread complex traits. They are looking for the genes that may predispose people to high blood pressure, heart disease, diabetes and most adult cancers. And on a far more incendiary scale, some researchers are seeking the genes that may put one at risk for a serious mental disease like manic-depressive illness, schizophrenia or alcoholism. They are looking for genes that influence sexual orientation, or the hunger for novel experiences, or the tendency toward introversion, or a taste for nicotine. And the only thing they are sure of in their various hunts is that no single gene can explain any of the behaviors they study, and that they will spend their professional lives qualifying every claim and cautioning audiences to please, please, please not overinterpret the results.

Nevertheless, after a long period of setbacks and missteps, the field of behavioral and psychiatric genetics is lately picking up steam and enthusiasm, reflected in the studies published and the commentary on them. With humorous didacticism, Dr. Eric Lander of the Whitehead Institute for Biomedical Research and the Massachusetts Institute of Technology in Cambridge, and Dr. Leonid Kruglyak of the Whitehead Institute warn their fellow biologists in a *Nature Genetics* editorial that while they may often greet statistics with "glazed-eyed indifference" (and you thought scientists liked that stuff!), they are going to have to knuckle under and take statistics more seriously in this new era of mapping complex traits.

Good statistics are essential on two counts. If a disease or trait has multiple causes, including many genes and that squishy business known as "the environment," then it will take a sound use of mathematics to find an interesting genetic connection in the first place. And once that connection has been identified, a rigorous statistical analysis will assure one that it is real before one rushes it into print.

"We wrote the commentary to try to put statistics in simple terms, so people can understand why we have to have a strict threshold before declaring linkage," Dr. Lander said. "We're going to see hundreds of papers on complex traits over the next few years, and we don't want people to be crying wolf."

In fact, the field of psychiatric genetics has only begun to emerge from a slump in which a number of wolves turned out to be dogs. In the late 1980s, several research teams reported finding genes for manic-depression and schizophrenia, which either were never confirmed by other researchers, or were proved wrong and had to be retracted.

But the incentive to move forward was too great to be deterred by a flub or two. For one thing, the illnesses are extremely common, each affecting about 1 percent of the population. They are devastating: for example, 30 percent of the hospital beds in the nation are occupied by schizophrenia patients, said Dr. Kenneth S. Kendler of the Medical College of Virginia in Richmond, an author with Dr. Richard E. Straub, a colleague there, of one of the schizophrenia papers. And scientists have had scant success trying to understand the illnesses through a nongenetic approach. "Psychiatry is still pretty much in the dark ages," said Dr. David Curtis of the Institute of Psychiatry in London, who contributed to the section on schizophrenia in *Nature Genetics*. "We have no idea about the basic biochemical abnormalities that occur in the course of the illnesses," and a genetic angle on the diseases may offer new insights into their cause.

Moreover, the evidence for a genetic contribution to these mental disorders is quite strong. Studies of twins, for example, showed a hereditary contribution of anywhere from 30 percent to 50 percent for schizophrenia, and somewhat more for manic-depressive illness. Researchers also said that the tools they had used before in searching for genes involved in complex mental disorders were too crude for the task.

"At the beginning of using linkage analysis to study schizophrenia, we used monogenic models," like those applied to the analysis of cystic fibrosis, said Dr. Hans W. Moises of Kiel University Hospital in Germany, an author on another of the reports. "These models are inappropriate where complex diseases are involved."

Recognizing the muddy history of their specialty, the psychiatric geneticists reporting on their discovery of an intriguing link to schizophrenia are resolutely cautious. What the groups have found is that there appears to be a gene on the upper arm of chromosome 6 (out of the 23 pairs of chromosomes all humans have) that may play a part in some unknown percentage of cases of the disease. But everybody involved admitted the evidence for chromosome 6p, as the region is known, is not overwhelming. Four teams found the association; and because independent replication is a benchmark for any scientific finding, the work is considered quite exciting. But in none of the individual studies was the result overwhelming, meaning that the chain of evidence is built of rather fragile links. Moreover, two other teams reported no connection at all between schizophrenia and 6p, and one of those scientists, Dr. Curtis, believes the finding will turn out to have been yet another false positive.

Dr. Ann E. Pulver of the Johns Hopkins University School of Medicine in Baltimore, who obtained one of those positive results, said that if the gene on chromosome 6 was involved in only a limited number of cases of the disease—say, 25 percent or less—there was no reason to expect every researcher to find it active in their sample of patients. She and others said the real test would come when researchers got their hands on the gene proper, rather than its approximate location, and then looked to see whether the gene is mutated in patients with schizophrenia.

That will take some doing. The 6p region of the chromosome contains many hundreds of genes. Just getting to that general neighborhood had been an extraordinary task. Dr. Kendler and his colleagues, who first identified the tantalizing chromosomal association, tried to make their work tenable by going to Ireland and collaborating with researchers in Dublin and Belfast. The Irish make a good study population for a number of reasons. They have large families, making it easier to trace genetic patterns and compare one relative with another. The Irish are more genetically similar than a comparable group of, say, Americans, and that makes it easier to home in on particular genes of interest. They are also more culturally homogenized, which means most people are exposed to relatively similar environmental conditions, an important consideration for a disease thought to have some environmental component. And finally, Dr. Kendler said, the Irish use very few recreational drugs beyond alcohol, which rules out the complicating factor of those drugs thought to induce a psychosis-like state.

The scientists identified 265 families with two or more people suffering from schizophrenia. They drew blood from as many of the 1,408 individuals as they could—patients and their nonafflicted relatives alike—and extracted DNA from the blood cells. Dr. Straub then screened the DNA with

200 so-called DNA markers, bits of radioactively tagged genetic material that serve as signposts indicating a location somewhere on the 23 chromosomes. He was looking for patterns of markers that would be found in the DNA of the schizophrenics but not in their nonafflicted relatives. That broadside approach showed chromosome 6 worthy of closer examination.

Dr. Straub then coordinated a collaborative effort with other teams known to be studying the genetics of schizophrenia. He told them about the tantalizing 6p connection, and asked them to scrutinize the DNA of their family groups to see if the link held up. In some cases, it did.

Other researchers have also detected promising links to chromosomes 8 and 22. But in all cases, they are far, far from singling out the genes themselves. Moreover, even if they isolate a gene and it proves to be involved in some cases of schizophrenia, it will never be as clean an association as the Huntington's gene is with Huntington's disease. Any one gene can only tell a minor part of the somber tale of madness.

An equally cautionary note applies to the recent report on a genetic link for male sexual orientation. In 1993, Dr. Dean H. Hamer of the National Cancer Institute in Bethesda, Maryland, and his coworkers reported that male homosexual brothers shared a part of the X chromosome far more often than would be expected by chance distribution alone. The finding suggested that a gene on this chromosome, which in a man is inherited from the mother, somehow in minor fashion influenced a man's choice of sexual partners at least some of the time. The new study confirms and extends the original discovery. This time, Dr. Hamer looked at the DNA not only of a fresh set of gay brothers, but also of their heterosexual brothers.

In addition, he checked the DNA of lesbian sisters, to see if the X chromosome might figure in female sexual orientation. He and his colleagues found that everything abided by their model. Again, gay brothers shared the chromosomal region more often than would occur by random chance alone, while their heterosexual siblings did not. Nor was there any association between the chromosomal region and female sexual orientation.

Dr. Hamer's work has yet to pass the ultimate trial of independent verification by another research team, however. A group in Ontario failed to see anything of interest on the X chromosome when it looked at the DNA of gay men, but Dr. Hamer and others said the two techniques used were too different to be comparable. Dr. Elliott Gershon of the National Institute of Mental Health is now recruiting sets of gay brothers to see if he can reproduce the findings, but the study is in its early stages. For now, the putative "gay gene," like the unicorn, remains a tantalizing quarry.

—NATALIE ANGIER, October 1995

Two Gene Discoveries Help Explain the Misfires of Epilepsy in the Brain

THE DISCOVERY of two new human genes that cause a form of epilepsy is expected to throw light on the nature of the disease as well as on the general principles by which the brain operates.

Epilepsy experts also hope that establishing a genetic basis for the mysterious and often alarming disease will help remove the stigma that often attaches to it. During epileptic seizures, patients' limbs may thrash about, frightening observers so much that in past centuries they assumed the body was possessed.

At least one form of the disease can now be attributed not to demons but to defects in genes that make a critical component of the electrical circuitry of brain cells. The defective component probably delays the recharging process in nerve cells that have fired off an electrical message to their neighbors. The delay presumably interferes with the brain's intricate timing pattern and somehow triggers the widespread lockstep firing of nerve cells that is the hallmark of an epileptic seizure.

"The discovery is tremendously exciting for not only people with epilepsy but all neuroscience—it's that timing behavior that is so crucial to the way a normal brain behaves," said Dr. Jeffrey L. Noebels, an epilepsy expert at the Baylor College of Medicine in Houston.

Epilepsy, said to affect 2 percent of people at one time or another in their lives, is a disease in which many different causes lead to the same symptoms, a derangement of the neuronal timing process. Some forms of epilepsy are caused by damage to the brain from infection or trauma, but 40 percent of cases are thought to be genetic in origin.

The two new genes have been detected by Dr. Mark Leppert and a team of geneticists at the University of Utah and other institutions. They found the genes by studying the pedigrees of several families who suffer from a rare but intriguing form of epilepsy. Their findings were published in the journal *Nature Genetics*.

The type of epilepsy is called benign familial neonatal convulsions because it runs in families and occurs in babies a few days old. The disease is termed benign because, in 84 percent of cases, it mysteriously disappears after a few weeks.

Possibly the nerve cells affected by the mutant genes cause their neighbors to take some action that reverses the effects of the defect after a few weeks of life. Another explanation, Dr. Leppert said, is that the gene in question is only used for a brief period in the infant's development, much like the gene that produces a special form of hemoglobin, known as fetal hemoglobin, for use in the womb.

The normal role of the new genes is to make critical components of nerve cells known as voltage-gated potassium channels. Nerve cells are able to transmit electrical signals because of channels, embedded in their membranes, that maintain an electrical imbalance between the nerve fiber and its surroundings. The channels allow sodium and potassium ions, which are atoms that have lost an electron, to cross the membrane in different directions.

A nerve signal is transmitted when the imbalance is reversed at the nerve's body, and the reversal propagates all the way down the nerve's fiber. The imbalance must quickly be restored for the cell to fire again, and the restoration is the role of the potassium channels, which open to let potassium ions rush across the membrane and restore the previous state of imbalance.

In the affected families, the genes that specify the design of the potassium channel protein have changes or deletions in the sequence of chemical letters that make up their DNA. These variations from the correct sequence presumably cause defects in the potassium channel proteins.

Dr. Leppert said that he first discovered a genetic link with the disease from studying a South Carolina family in 1989. It has taken until now to pinpoint the gene among the 70,000 others contained by human chromosomes and to analyze its differences from the normal form. An epilepsy-associated gene was discovered in 1996 in Finland, but the gene causes general nerve degeneration. The new genes are the first pure epilepsy-causing genes to be found in humans, Dr. Leppert said.

Dr. Noebels said he believes that perhaps a hundred other genes that can cause epilepsy when damaged remain to be discovered.

For those affected by epilepsy, "the genetics has done two things," Dr. Noebels said. "People suddenly realize epilepsy is not the devil striking you dead but a genetically caused disease. That makes the disease seem more mainstream, though it is still scary."

The other advantage offered by the genetic dissection of epilepsy is that when a genetic defect is known, patients can be given an accurate diagnosis along with information about which family members are affected and which are not.

Most people with epilepsy are fine between seizures and for them the worst part of the disease is its unpredictability.

"All of their behavior is constrained because they fear to go swimming or driving," Dr. Noebels said. "Society is inordinately afraid of it, since any one of us could have a heart attack at any time."

Dr. Leppert believes that other forms of epilepsy are likely to be traced to defects in other kinds of membrane channel proteins. Even the roundworm, a standard laboratory organism, has more than 50 genes specifying different types of potassium channel protein, so it is likely that humans possess even more. Identification of these genes through their ability to cause epilepsy should help to explain what they do in the human brain and why an organism should need so many different kinds of channels.

—NICHOLAS WADE, December 1997

Studies Outline Clever Tricks of Viruses

AFTER A WAR has been waged for many years, ingenious tricks and countermeasures and counter-countermeasures emerge on both sides. When a war has gone on for millions of years, like that between viruses and the human immune system, the stratagems can attain a high level of refinement.

Viruses have learned to hide inside cells, and the immune system has developed a system for identifying cells that hold hidden invaders. Viruses have learned how to deceive the surveillance system; the immune system has a backup method for marking the cells where it was deceived.

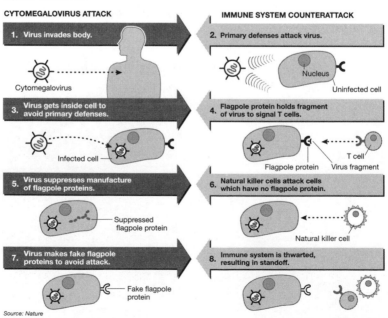

Move and Countermove

Viruses and the immune system fight a sophisticated war with many levels of attack and defense. Several of these moves have now been worked out for a virus known as cytomegalovirus. When the virus replicates inside a human cell, the cell displays pieces of the virus on protein "flagpoles." Immune system T cells recognize the viral parts and kill the infected cells. But cytomegalovirus can suppress the flagpoles, so a second kind of immune defender, natural killer cells, destroys cells with no flagpoles. Cytomegalovirus decoys the killer cells by erecting fake flagpoles on its host cell.

CYTOMEGALOVIRUS ATTACK

1. Virus invades body.

Cytomegalovirus

3. Virus gets inside cell to avoid primary defenses.

Infected cell

5. Virus suppresses manufacture of flagpole proteins.

Suppressed flagpole protein

7. Virus makes fake flagpole proteins to avoid attack.

Fake flagpole protein

IMMUNE SYSTEM COUNTERATTACK

2. Primary defenses attack virus.

Nucleus
Uninfected cell

4. Flagpole protein holds fragment of virus to signal T cells.

T cell
Flagpole protein   Virus fragment

6. Natural killer cells attack cells which have no flagpole protein.

Natural killer cell

8. Immune system is thwarted, resulting in standoff.

Source: Nature

N.Y. Times News Service

220

This hide-and-seek game is of some consequence for the body's health, since tumor cells, which also must dodge the immune system's scrutiny, use several of the same tricks as viruses.

Biologists do not yet fully understand the multiple layers of interaction between the body and the viruses that learned how to colonize it. But though much remains to be learned, several stages of ploy and counterploy can now be described for a particularly crafty microbe known as cytomegalovirus, whose latest trick was described in the journal *Nature*.

A cytomegalovirus that has burrowed into a cell must first cope with the immune system's method for uncovering infiltrators. Cells make what can be thought of as flagpole proteins, which jut out through the cell's membrane, and every individual makes a characteristic brand. Usually the flagpole proteins signal that the cells are the body's own, an advertisement of health and selfhood.

When a virus starts to replicate inside a cell, the flagpole proteins start to display fragments of the virus. The flagpoles are constantly inspected by the immune system's patrolling enforcers of purity, known as killer T cells. If the T cells sense a foreign protein on the flagpoles (known to immunologists as major histocompatibility complex class 1 proteins), they order the immediate execution of the stricken cell for the public good.

Cytomegalovirus's first subterfuge is to jam this red-flag mechanism. It has developed several ways of doing this, one of which is to bottle up the flagpole proteins in their assembly compartment and gag the cell's cry for help.

Clever, but the immune system has evolved a countermeasure. With the T cells blinded to danger, another, less discriminating weapon was developed, known scientifically as an NK cell. (NK stands only for natural killer, but immunologists like acronyms.) The NK cells are the berserkers of the immune system. Their orders are to ax-murder any cell that displays no flagpoles.

So the immune system had it all worked out: T cells to kill infected body cells with flagpoles bearing virus body parts, NK cells to kill cells without flagpoles.

But cytomegalovirus has a ruse to evade NK cells, too. It has either developed on its own, or stolen from human cells, a gene that makes flagpole proteins. By erecting fake flagpoles on the cells it has infected, the virus protects them and itself from NK attack. Discovery of the false-flag ruse was reported in *Nature* by Dr. Jack L. Strominger and his colleagues at Harvard University, who studied human cells, and by Dr. Helen E. Farrell and her colleagues at the University of Western Australia, who noticed the same trick in infected mice.

Dr. Strominger said he got the idea that cytomegalovirus might be decoying the NK cells after finding a similar mechanism in cells of the human fetus. Fetal cells must suppress their flagpole proteins to avoid being perceived as foreign by the mother's immune system, which would reject them, so they need a false flag to avert assault by the mother's NK cells.

"Different viruses have evolved different ways of getting rid of expression of class one proteins," Dr. Strominger said, referring to the flagpole proteins. "But the mechanisms are amazingly different in different viruses."

As long as its defenses are intact, the immune system somehow keeps cytomegalovirus well in check despite the microbe's evasion of both its T cell and NK defense mechanisms. The virus, which is common and generally harmless, is forced to bide its time until the immune system is changed or compromised. Then it will strike. It can cause blindness in AIDS patients and birth defects or death in the fetus.

"Viruses teach us a lot of immunology," said Dr. Lynn W. Enquist, a virologist at Princeton University. "With acute viruses like flu, it's wham, bam, thank you ma'am—the immune system is very good at taking care of them." But slower-growing viruses need ways to cope with the immune system. "These big viruses like HIV and hepatitis have to figure out how to replicate in the presence of the immune system," he said. "They have all learned how to do this, and they do it in fifty-seven different ways." Adenovirus, the cause of the common cold, is another microbe with a bag of cunning tricks. Like cytomegalovirus, it has genes that suppress the telltale flagpole proteins. To get itself replicated, adenovirus needs another fancy stratagem. It must get its host cell to divide, something that cells are allowed to do only at the body's express command, since otherwise chaos and cancer would ensue.

The official restraint on division is exercised inside many kinds of cells by a master protein called Rb. Adenovirus has a genetic tool for knocking out Rb and tripping the cells into division. Human cells have a desperate countermeasure. All are equipped with a suicide program that can be activated if a breach in the integrity of the cell's genetic data systems is sensed.

When the cell's monitors detect that an adenovirus is replicating, they trigger the suicide routine to destroy both themselves and the invader.

The adenovirus has a riposte for this measure, too: a gene that can jam the self-destruct program.

Curiously, the cell mechanisms that adenovirus disrupts—the flagpoles, the Rb growth-constraint system and self-destruct program—are the same that are sabotaged by tumor cells.

"It's quite amazing," said Dr. Frank McCormick of the University of California at San Francisco. "All the things that tumors do are the very same things the virus does."

Dr. McCormick has an ingenious scheme for pitting the adenovirus against human tumors. The plan, now being tested by Onyx Pharmaceuticals of Richmond, California, depends on a strain of adenovirus that has lost the gene for knocking out the cell's self-destruct mechanism. The strain therefore cannot grow in normal cells, but Dr. McCormick guessed it should grow in tumor cells, because they have inactivated their suicide machines, and the virus should, with luck, destroy them. Onyx's Phase 1 trials, designed to test safety, have shown no toxicity so far and the trials are being expanded.

The AIDS virus has the deadliest trick of all, that of infecting the cells of the immune system itself. HIV's targets are the helper T cells. Cousins of the killer T's, which execute infected cells, the helper T cells energize other cells to respond to infection. "When you infect T lymphocytes, you are infecting the officer corps," said Dr. Wade P. Parks, a virologist at New York University.

Like other viruses that plan on staying around in the body, HIV has to disable the flagpole alerting system. It has a particularly cunning method. One of HIV's genes, called nef, makes an agent that catches on to the flagpole proteins, to the base of the pole that is planted inside the cell's membrane.

The other end of the nef protein carries an address label, readable by the cell's internal sorting system that directs proteins to their proper place.

The message carried by the nef protein attached to the base of the flagpoles says, in the cell's sorting code, "Haul to garbage dump and recycle," tricking the infected cell into pulling down its flagpole proteins and destroying them. This role of the nef gene was reported by Dr. Didier Trono of the Salk Institute and his colleagues in the journal *Immunity*.

The nef agent also tags the CD4 proteins for destruction in the same way. Like the flagpole proteins, the CD4's stick up through the cell membrane. Their role is to help the T cell recognize the other immune cells it activates, but they themselves are recognized by the AIDS virus as its portal for infecting the cell.

What does HIV gain from having the cell destroy CD4 proteins? "That's the million-dollar question," Dr. Trono said. The reason may be to prevent other viruses from entering the cell or because in latching on to the CD4 proteins, the nef agent dislodges and energizes another protein that is known to activate the T cells. The cell's activation may be what HIV seeks, Dr. Trono suggested, with the destruction of the CD4 proteins being a by-product.

For an AIDS virus to pull down the flagpole proteins in a cell it has infected presumably renders the host cell vulnerable to assault by NK cells.

"So it is reasonable to ask if the virus has a trick to prevent NK attack, too," Dr. Trono said.

—NICHOLAS WADE, May 1997

7

GENES AND LIFE SPAN

Many people assume that aging is an inevitable wearing out of the body's tissues, much like a car whose parts wear out after their designed lifetime has been exceeded. Evolutionary biologists think of aging as the result of inevitable neglect: Once an organism has passed its reproductive years, the shaping forces of evolution no longer care what happens to it and so have no reason to prolong life span beyond that point.

These two gloomy viewpoints may both be wrong. Living cells have the capacity to renew themselves constantly; any analogy to a mechanical device is superficial. And evolutionary biologists have seen their theory put in question by the discovery of genes that extend life span in the roundworm.

A dramatic new finding is that the life span of human cells cultured in the test tube can be extended indefinitely. Usually cultured cells will divide a finite number of times and then die. It now seems their life span is measured by a mechanism that counts the number of times they have divided. The mechanism resides in a special region of DNA at the tips of the chromosomes, known as the telomeres. The telomeres get shorter at each division and when they reach a certain minimum length, the cell is forced into senescence.

It is tempting to suppose that many tissues of the body are under the same control and wear out only when their telomeres have reached the minimum length. Since telomeres and life span can now be lengthened for human cells in the test tube, can the same be done for cells in the body?

The implications of the telomere mechanism are still being worked out. They offer, for the first time, a scientific basis for hoping that senescence may be amenable to treatment.

Can Life Span Be Extended?
Biologists Offer Some Hope

THOUGH IT IS OFTEN SAID that only two things in life are certain, maybe only one is truly unalterable: taxes. Biologists who study aging are beginning to talk of death as if it was not entirely immutable.

Until recently, the subject of senescence has been dominated by gloomy findings about limits and genetic fixity. Actuaries have believed that the likelihood of death increases every year that a person lives. Evolutionary biologists teach that natural selection culls out bad genes that threaten youth but has no reason to help prolong a person's life after the time of child rearing. Biologists know that human cells grown in a test tube will divide up to 50 times and then fail, as if some finite thread of life had run out.

None of these dismal facts has been disproved, but new developments have made them seem suddenly a little less ironclad. Gerontologists were surprised to learn recently, from surveys of aging populations in 13 Western European countries and Japan, that human mortality rates do not accelerate throughout the life span, as was supposed. The chance of dying starts to decelerate around age 80 and levels off sharply after 110. As if to emphasize that mortality rates ease after 110, in 1997 Jeanne Calment of France died at the age of 122, the longest human life span that is well attested.

Biologists studying aging at the level of cells and genes, meanwhile, are finding that life span is far easier to manipulate than they had expected, at least in certain laboratory organisms. Some eight genes have now been found that can lengthen the life span of the laboratory roundworm, *C. elegans*, in one instance by fourfold. That is equivalent to making a person live 340 years.

And biologists at the Massachusetts Institute of Technology recently announced that they had discovered a basic mechanism that causes yeast

The Roots of Mortality

Many changes take place in the body's tissues during aging. Evolutionary biologists have said no single gene or fundamental mechanism is likely to be responsible for aging because it involves a multitude of changes. But a basic mechanism of aging has now been found in normal yeast. This aging mechanism can be triggered early by a gene that has a counterpart in humans. Defects in the human version cause Werner's syndrome, in which many aging signs appear prematurely.

The Relentless March Of Time

HAIR AND NAILS
Hair often turns gray and thins out. Men may go bald. Fingernails can thicken.

BRAIN
The brain shrinks, but it is not known if that affects mental functions.

THE SENSES
The sensitivity of hearing, sight, taste and smell can all decline with age.

SKIN
Wrinkles occur as the skin thins and the underlying fat shrinks, and age spots often crop up.

GLANDS AND HORMONES
Levels of many hormones drop, or the body becomes less responsive to them.

IMMUNE SYSTEM
The body becomes less able to resist some pathogens.

LUNGS
It doesn't just seem harder to climb those stairs; lung capacity drops.

HEART AND BLOOD VESSELS
Cardiovascular problems become more common.

MUSCLES
Strength usually peaks in the 20's, then declines.

KIDNEY AND URINARY TRACT
The kidneys become less efficient. The bladder can't hold as much, so urination is more frequent.

DIGESTIVE SYSTEM
Digestion slows down as the secretion of digestive enzymes decreases.

REPRODUCTIVE SYSTEM
Women go through menopause, and testosterone levels drop for men.

BONES AND JOINTS
Wear and tear can lead to arthritic joints, and osteoporosis is common, especially in women.

Time's Metronome

The genetic material called ribosomal DNA seems to hold the key to aging in yeast.

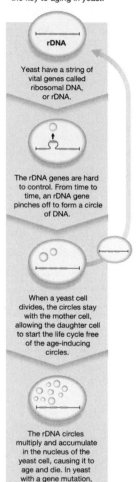

rDNA

Yeast have a string of vital genes called ribosomal DNA, or rDNA.

The rDNA genes are hard to control. From time to time, an rDNA gene pinches off to form a circle of DNA.

When a yeast cell divides, the circles stay with the mother cell, allowing the daughter cell to start the life cycle free of the age-inducing circles.

The rDNA circles multiply and accumulate in the nucleus of the yeast cell, causing it to age and die. In yeast with a gene mutation, the aging process is speeded up.

Source: Dr. Leonard Guarente/Massachusetts Institute of Technology.

cells to age. The mechanism can be triggered by failures in a gene whose counterpart in humans causes Werner's syndrome, a disease with symptoms that strongly resemble premature aging.

Roundworms and even yeast are much closer to humans genetically than their outward forms suggest. But similarities aside, the importance of the new findings is that they seem to contradict in several ways the long-standing theory of aging developed by evolutionary biologists. Their theory says there is no basic mechanism of aging, just a lot of things that go wrong at about the same time. They believe aging is controlled by a multitude of genes, with no single gene having any big effect, and hence that it is folly to think of senescence as a simple disease susceptible to some single intervention.

"Ten years ago, evolutionary biologists had this field to themselves, and it has really been invaded by molecular biologists," said Dr. Leonard Guarente, who reported the finding about aging in yeast. "The molecular biologists really believe they will be able to identify the underlying mechanisms of aging, and the evolutionary biologists believe this isn't going to work." If basic cellular mechanisms of aging can be identified, there would be at least a chance of trying to thwart them.

"The field has gotten very hot the last few years," said Dr. Thomas E. Johnson, a biologist at the University of Colorado who studies senescence. "The identification of genes that slow the rate of aging has made people recognize that the fundamental aging process, even in higher organisms, could be under genetic control and can maybe be modulated."

An interesting collision thus seems to be shaping up between evolutionary biologists, who believe, for widely accepted but largely theoretical reasons, that fountains of youth cannot exist, and molecular biologists, who regard the question as an open issue to be settled by experiment.

The logic of the evolutionary biologists is hard to fault. Animals are not designed to live forever, they say, because there would be no point. They quickly get killed by predators or accidents, which is why so few elderly animals are seen in nature. Better to channel resources into breeding prolifically and early in life. It is not that evolution, meaning the force of natural selection, cannot design organisms of immense longevity: The bristlecone pine lives for 5,000 years. But for most creatures, the necessary trade-off between fertility and longevity yields a much briefer life span. The Garden of Eden account said it first: Sex and death are tightly linked.

One mechanism that brings this about, evolutionary biologists believe, is that the force of natural selection decreases at later stages of life span, which means it is unable to favor genes that confer benefits only in older individuals. Also, among the many genes that have more than one effect on the body, those that are good for youth and bad for age become incorporated in the population because their holders have many descendants; the opposite is true of the genes that are rough on teenagers and kind to oldsters.

The major elements of this theory were contributed by the biologists P. B. Medawar, W. D. Hamilton and George C. Williams, and it has remained unchallenged since Dr. Williams's formulation of it 40 years ago. Asked about the gloominess of its conclusions on the fixity of life span, Dr. Williams, who is at the State University of New York at Stony Brook, said it was gloomy, too, when alchemists found they could not turn lead into gold. "People who think they are going to find a fountain of youth, whether at the molecular or any other level, are not going to be successful," Dr. Williams said.

The discovery of single genes that affect life span has come as a surprise to evolutionary biologists. People with Werner's syndrome exhibit many of the complex symptoms of aging at a young age. Yet all were found in 1996 to be caused by a single gene.

By mutating certain genes, biologists can make Methuselahs out of the roundworm, an animal that shares many genetic mechanisms with humans. These long-lived worms generate "something of a headache for evolutionary biologists," Dr. Linda Partridge of the University of Edinburgh wrote in the journal *Nature,* "because evolutionary theories of aging strongly predict that no single mutation should be able to have such a dramatic effect."

At a 1996 aging symposium convened by the National Academy of Sciences, Dr. Partridge said, "Everything we know about the evolution of aging suggests that it is probably the most polygenic of traits," meaning that many genes control it. "Fertility and viability are affected by all the genes in the genome," she said. "The prospects for genetic intervention, therefore, must be explored with some circumspection."

That was the prevailing view in biology seven years ago when Dr. Guarente of MIT started his work on aging in yeast. He hoped to find a single mechanism that made cells age, despite the theorists' prediction of many ailments. "I understood their argument and found it reasonable, but with a biological problem as important as aging, it's worth taking a gamble," he said.

He and two graduate students decided to invest a year looking for genes that favored longevity in yeast. After a year, they had gotten nowhere but decided to keep going anyway. It took four more years to find the first clue, a set of genes that somehow fostered longevity when their protein products migrated to the nucleolus. This is a compartment within the nucleus of the cell reserved for the manufacture of ribosomes, the cell's machine tools for making proteins.

In December 1997, Dr. Guarente and a colleague, Dr. David A. Sinclair, reported in the journal *Cell* that they had found out what goes wrong in the nucleolus, and that they believe the defect is a fundamental cause of aging in yeast. Snaking through the nucleolus is a chromosome that carries the genes for making ribosomes. To help the cell make ribosomes quickly, one of these ribosomal genes exists in a string of multiple copies.

The defect arises because the cell's machinery for handling chromosome recombination gets overloaded when it has to deal with repeated copies of the same gene. Every so often, one of the ribosomal genes accidentally pinches off from its chromosome and forms a circle of DNA.

It so happens that each ribosomal gene possesses a signal that orders its own replication. So once a circle has pinched off from its chromosome, it multiplies freely, eventually clogging up the nucleolus until the little assembly room can no longer make its ribosomes and the cell's protein production falters. A persuasive feature of this explanation is that when the yeast cell divides, the cloying circles all remain with the mother cell; the daughter cell begins to age only when it accidentally creates a ribosomal DNA circle.

Dr. Guarente found the Werner's syndrome link in a quite separate experiment. After discovery of the human gene that causes Werner's syndrome, yeast was found to possess a counterpart gene with a similar DNA sequence. Dr. Guarente deleted the yeast's copy of its Werner's-like gene to see what would happen. To his surprise, the gene triggered the same accumulation of circles that he had seen in normally aging yeast cells, except that it caused them to appear early in the yeast's life, just as aging occurs prematurely in Werner's syndrome.

Werner's syndrome experts caution that the disorder is not necessarily the same process as human aging, despite striking similarities. But the behavior of the yeast version of the Werner gene is an interesting clue that cells in other species besides yeast may age because of damage to the nucleolus. Dr. Guarente is now looking for circles in the nucleoli of mouse and human cells.

Dr. George M. Martin, a leading expert on Werner's syndrome and aging at the University of Washington in Seattle, praised Dr. Guarente's work but noted that humans, unlike yeast, have two kinds of cells, those that divide often during life, like skin and intestinal cells, and those that never divide after maturity, like brain and muscle cells. The yeast mechanism, even if found to occur in the dividing type of human cells, "wouldn't tell us about the mechanism of aging of neurons and muscle cells," he said.

It is here that studies of the roundworm may be pertinent because the grown animal is composed of cells that never divide again. In November 1997, Dr. Cynthia Kenyon, who studies roundworm aging at the University of California at San Francisco, reported a gene that in its mutated form doubles the worm's life without affecting its fertility, an apparent defiance of the evolutionary biologists' trade-off rule.

"At an age when the normal worms look horrible, the mutant worms look just fine—like the difference between being in a nursing home and on the tennis court," she said. If the same kind of genetic switch exists in other species, "it would be interesting to know if we could affect the life span of higher organisms," she said.

Machines wear out, but biological systems have great capacity for self-repair. What is the inherent longevity of living cells? Do they just fall apart, or could they live indefinitely if some genetic program did not close them down? Could evolution's programming be rewritten? Cancer cells grown in the test tube last so long they are called immortal. Maybe healthy cells divide only 50 times in the laboratory simply because test-tube life does not agree with them. If so, that often-cited limit means less than it would appear. Germline cells, the egg and sperm, are in a sense immortal because they transcend the generations. Dr. Guarente believes that life span may be imposed on top of an inherent robustness that living cells have inherited from primeval ancestors.

The evolutionary biologists are unlikely to be wrong in broad principle. But they have yet to account for the new results that are flying their way, like the deceleration of human aging and the single genes that postpone senescence in laboratory organisms. The molecular biologists, meanwhile, are gathering confidence to challenge the dismal view that life span is immutable.

"Once the Achilles' heel(s) of the cell has been precisely identified, one can imagine interventions to slow changes that occur with aging," Dr. Guarente wrote in 1997 in the journal *Genes & Development.* "The spirit of Ponce de León lives on!" It has been a long time since a serious biologist dared endorse such an idea.

—NICHOLAS WADE, January 1998

A Clue to Longevity Found at Chromosome Tip

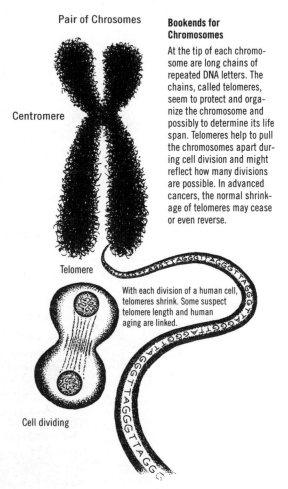

Pair of Chrosomes

Centromere

Telomere

Cell dividing

Bookends for Chromosomes

At the tip of each chromosome are long chains of repeated DNA letters. The chains, called telomeres, seem to protect and organize the chromosome and possibly to determine its life span. Telomeres help to pull the chromosomes apart during cell division and might reflect how many divisions are possible. In advanced cancers, the normal shrinkage of telomeres may cease or even reverse.

With each division of a human cell, telomeres shrink. Some suspect telomere length and human aging are linked.

Patricia J. Wynne

HUMAN CHROMOSOMES, shaped like cinch-waisted sausages and sequestered in nearly every cell of the body, are famed as the place where human genes reside. But now some scientists are finding that a few small architectural details of the chromosomes are at least as tantalizing as the 100,000 genes they harbor.

Biologists have discovered that at the very tips of the chromosomes are extraordinary structures built of six DNA letters repeated over and over, thousands of times, like monotonous molecular chants. They are learning that these structures, called telomeres, are profoundly important at every stage of the cell's life, protecting the chromosomes against harm, organizing them into their proper position and, most provocatively, possibly serving as a sort of timekeeping mechanism that tells the cell how old it is.

The work has broad implications for the study of aging and for understanding the perverse sort of cell immortality that is manifested by cancerous cells.

Scientists have determined that human telomeres are shaved down slightly whenever a cell divides, and they believe the length of the chromosome tip could offer the cell a measure of how many times it has divided— and how many divisions remain before the cell's innate life span is spent. And though individual cells die and are replaced throughout a person's life, some researchers suspect that there is a strong correlation between telomere length and the aging of the entire human being. On average, the telomeres of a 70-year-old are much shorter than those of a child, and biologists propose that once the telomeres of most cells in the body fall below a certain length, death is likely to follow.

The results could point the way toward developing new treatments to help shore up elderly cells, particularly in parts of the body where wear and tear are most relentless, such as along the coronary arteries. But scientists emphasize that many components of the body change with time, and that the puzzle of aging and death is surely more complex than telomere status alone.

Other studies have shown that telomeres undergo a dramatic transformation during the genesis and progression of cancer. When a cell becomes malignant and begins dividing out of bounds, its telomeres at first become shortened, whittled down with each illicit cellular split. Presenting their results in San Diego at a meeting of the American Association for Cancer Research, several groups of scientists said that telomere length could be used to determine the stage of a cancer cell: the more diminutive the tip compared with the patient's healthy cells, the more advanced the malignancy.

More intriguing still, new experiments suggest that at a very advanced and highly aggressive stage of tumor development, the telomere shrinking may cease, and may even reverse itself. Reporting in the European molecular biology journal *EMBO,* Dr. Carol W. Greider of Cold Spring Harbor Laboratory in New York, Dr. Silvia Bacchetti of McMaster University in Hamilton, Ontario, and their colleagues have found that most cancer-like cells growing in the laboratory eventually die, but that a fraction of those cells will become immortalized, essentially able to divide forever.

The scientists have found that in those immortalized cells an enzyme called telomerase, which is normally quiet in adult cells, has swung into action and has started rebuilding the telomeres, a skill that may prove deadly for a cancer patient. If telomere length is the timepiece designed to tell a cell to expire gracefully, and if that length fails to dwindle from one doubling to the next, no internal signal may remain to prevent the cell's ambitions from overwhelming the entire body.

But the new insights into telomerase also suggest a way of blocking late-stage cancer. The researchers propose that if a drug could interfere with the activity of the telomerase enzyme, the immortal cancer cells could be destroyed without damaging most normal cells. Searching through a variety of healthy tissue types, the biologists so far have been able to detect hints of telomerase activity only in sperm cells, where the enzyme seems necessary to keep chromosomes at the most youthfully elongated length possible. Thus, an attack targeted at telomerase may have no more severe effect than to cut back sperm production.

Designing telomerase blockers may not be difficult. Other work suggests that the enzyme somewhat resembles reverse transcriptase, the molecule at work in the virus that causes AIDS, and biologists are testing drugs known to block reverse transcriptase to see if they are effective against telomerase and hence are a potential remedy for advanced cancers.

Many basic researchers are passionate about telomeres for what they believe the structures will reveal about the fundamental nature of chromosome design. Regions of the genetic material that lie adjacent to the telomeres seem to be particularly prone to recombining, able to mix and match with other chromosomes during the creation of eggs and sperm cells. Because they shuffle around so much, these chromosomal regions, in theory, are more likely than other zones of the chromosomes to mutate in ways that could start cancer tumbling. Scientists are seeking to determine whether the telomeres might protect the genes that directly abut them, perhaps by attracting the zealous attention of the cell's repair enzymes, and in so doing whether they help counterbalance the risk that comes with living next door to a telomere.

"There's no end to the interest of the scientific community" in the study of telomeres, said Dr. Elizabeth H. Blackburn of the University of California at San Francisco, described by her colleagues as the queen of telomeres. "They're turning out to be quite intriguing, and to be made in such a funny way."

That intrigue began in the early 1970s, when one of the greats of modern genetics, Dr. Barbara McClintock of Cold Spring Harbor, Long Island, observed during her studies of corn that broken chromosomes were extremely unstable, and she proposed that normal chromosomes must possess protective structures to prevent their degradation. Those structures turned out to be the telomeres.

Dr. Blackburn's laboratory took up the detailed investigation of these chromosomal rail guards, turning to single-cell organisms similar to the paramecia that swim through pond water. Such creatures are equipped with tens of thousands of extremely tiny chromosomes, rather than the mere 23 matched pairs found in human cells. More chromosomes per cell means many more telomeres that can be isolated and deciphered.

Dr. Blackburn and her student, Dr. Greider, and others described the peculiar telomeres in great detail, finding that they were composed of a sequence of only six DNA bases, or molecular subunits, repeated thousands of times at both ends of every single chromosome. In human cells, the sequence is TTAGGG, and it is repeated anywhere from 1,000 to 3,000 times, depending on the cell type and its age. The sequence does not mean anything on its own, but it serves, among other tasks, to stabilize the chromosomes.

"The telomeres are like bookends," said Dr. Blackburn. "They hold everything in place."

Dr. Greider then found how those telomeres are created and maintained, particularly in organisms like the pond creatures and yeast cells, which divide indefinitely. She discovered the enzyme, telomerase, and found that it had an outstanding property, observed in no other enzyme. Within the enzyme complex is a little template made of the chemical RNA, a sort of rubber stamp of the six necessary letters that can be used to add new sequences to the chromosome ends whenever they are required.

"The enzyme carries this template sequence along with itself," she said. "This was something completely new that had never been seen before."

Researchers have since observed that while the telomerase rubber-stamp enzyme works in paramecia and other simple species full-time, it does not seem to be active in most human tissue, or in that of other mammals. Instead, mammalian telomeres seem to shrink with age, a result of the way the linear chromosomes are duplicated whenever a cell divides.

Normal human cells divide between 70 and 100 times, and with each split they lose about 50 letters of their telomeres from both chromosomal ends. Before the telomeres vanish altogether, the chromosomes sandwiched between them become warped and sticky, attaching themselves to other chromosomes in such twisted configurations that the cell eventually dies.

Scientists are still trying to figure out whether the gradual loss of telomeres actually causes the cell to age and die, or is merely an incidental by-product of the passage of time.

Some believe that telomere degradation is no more responsible for the aging of the cell than graying hair is for the aging of an entire human, and they await evidence to the contrary.

Dr. Calvin B. Harley of McMaster University, who collaborates with Dr. Greider, believes that such proof is forthcoming. In studies of cells proliferating in laboratory dishes, he and others have found that the shrinking telomeres keep the chromosomes remarkably stable until right before the cell's allotted time is through, rather than everything falling apart in a gradual, piecemeal fashion. He suggests this is no coincidence but that the extremely short telomeres could somehow signal other molecules in the cell to stop cell division in preparation for death.

"The genetic stability of the chromosomes remains very normal until the cells get very close to the end of their life span," he said. "At that point, there's a dramatic increase in chromosomal aberrations. This suggests a functional purpose for the telomeres" as the cell's clock.

But the link between chromosome tips and aging remains slippery, he said, and no miraculous telomeric antiaging potion is likely to emerge from the research any time soon.

More therapeutically promising are studies of telomeres and cancer. Beyond the possibility of blocking the aberrant telomerase enzyme at a late stage of cancer, researchers also believe there are applications of telomere research to treating early phases of a tumor. Dr. Bradford Windle of the Cancer Therapy and Research Center in San Antonio, Texas, is working on an approach he calls chromosome healing, to help repair chromosomes that are splintered through mutational blows by radiation or other environmental assaults.

As scientists have seen, a broken chromosome is likely to recombine mistakenly with other chromosome bits it is not meant to encounter. That sort of degradation usually leads to cell death, but every so often a misguided chromosomal fragment rearranges itself in a way that switches on a cancer gene and a tumor is born.

Working with hamster cells, Dr. Windle and his colleagues have explored whether or not artificial telomeres can be applied to broken chromosomes to prevent their perilous entanglement with the wrong sort of genetic partner. They have found that by flooding the cells with the magic sequence of molecular letters, TTAGG, they can cap off the snapped pieces and thereby forestall any perilous chromosomal rearrangements. Dr. Windle suggests that the approach could be most useful in those people with a predisposition to cancer, whose chromosomes may have a tendency to break and who therefore are in need of molecular healing.

Whether the approach will work in human patients remains to be demonstrated. For now, the study of telomeres offers yet another graphic demonstration of how cell death and malignant growth are closely intertwined, and that sooner or later, if one doesn't get you, the other one will.

—NATALIE ANGIER, June 1992

Cells' Lives Stretched in Lab

IN A FINDING that suggests the thread of life can be extended, at least for human cells, biologists have succeeded for the first time in coaxing cells grown in a test tube to live far beyond the limit at which they usually fail.

The finding may help answer the question of why human cells age, and may also offer a practical means of treating certain diseases by rejuvenating aging cells.

Despite some scientists' euphoric talk about the discovery of a cellular "fountain of youth," others cautioned that the cell senescence mechanism is also a principal defense against cancer, and that bypassing it could be dangerous.

The finding makes use of a newly discovered human gene that resets an inherent limit imposed by nature on the number of times a human cell can divide. Human cells grown in a test tube will usually divide about 50 times. But cells that have undergone the new treatment have bypassed this limit and have divided 90 times with no sign of abnormality.

The result has broad scientific and medical interest because it elucidates a mechanism in living cells that is central both to aging and to cancer. Although the mechanism, known as telomere shortening, has long been suspected to work this way, proof has been unattainable until now.

The Geron Corporation of Menlo Park, California, which owns or has applied for several patents on the gene, known as the telomerase gene, says its technology will rejuvenate cells involved in age-related diseases and help diagnose and treat cancer.

"These cells have an indefinite life span as far as you can tell," said Dr. Calvin B. Harley, an author of the finding and vice president of Geron.

Another author of the study, Dr. Jerry W. Shay, of the University of Texas Southwestern Medical Center, said the finding would "now allow us to take a person's own cells, manipulate and rejuvenate them and give them back to the same patient." The rejuvenated cells could help grow new skin for burn victims and aid people with diseases caused by the failure of aging cells to divide, like macular degeneration.

The new research was conducted by two teams of scientists led by Dr. Harley at Geron and by Dr. Shay and Dr. Woodring E. Wright of the Texas medical center. Their work was reported in the journal *Science*.

Disclosure of the findings sent Geron's stock up 44 percent on the Nasdaq stock exchange, where it closed at $14.375 a share.

Experts cautioned that as important as the new results were, there was a Catch-22 in the way evolution had designed the telomerase system. The mechanism almost certainly evolved, at least in humans, as a way to limit cancer. Hence any medical use of rejuvenated cells in which the division limit has been subverted may weaken one of the body's salient defenses against tumors.

"Geron would have us believe that telomerase is the key to immortal life, and I have no idea if there is any wisp of truth in that," said Dr. Robert Weinberg, a leading expert on cancer genetics at the Whitehead Institute in Boston.

The mechanism at the heart of the new work is a system developed by certain living cells for managing their chromosomes, the rod-shaped structures in which the cell packages its genetic programming tapes. The device that copies the long double helix of DNA when the cell divides has a peculiar defect: It cannot copy the last few units of DNA at the chromosome's tip. So each time a cell divides, its chromosomes get a little shorter. The end section of the chromosomes, known as the telomere, thus resembles some thread of life that is used as the tally for counting off cell divisions. When it runs out, that means that the cell's ability to divide is gone.

This tempting analogy became much strengthened when biologists found that all human cells possess a gene whose product, named telomerase, can lengthen the telomeres. In normal cells, the gene is firmly silenced. But in human germline cells, the egg and sperm, telomerase is active and maintains the telomeres at a constant length—some 15,000 chemical letters of DNA. It is also active in cancer cells, which have learned to switch on telomerase and bypass a mechanism that would otherwise shut them down.

Skeptics of this theory began to question whether the telomerase shortening was really as important as advertised. In 1997 Dr. Carol W. Greider of

Johns Hopkins University engineered a breed of mice that lacked a telo-merase gene. The mice are alive and well after several generations, and their cells show no particular cancerous tendencies.

"The theory came under tremendous skepticism," said Dr. Harley, who was one of its earliest proponents. "People said it was gibberish or based on theology."

What use can be made of the potentially awesome power to rejuvenate human cells?

Geron and scientists associated with the company believe that many diseases of aging can be treated by removing cells from a patient's body, restoring their telomeres and replacing the cells. As examples they cite aging of the skin, macular degeneration—or retinal decay, a leading and incurable cause of blindness—and atherosclerosis, or hardening of the arteries. An advance on any one of these would be a notable success.

The technique, the scientists say, could also be used to grow new skin for burn victims from their own cells. Farther ahead lies the possibility of developing drugs to shut telomerase down in the many kinds of cancer cells that activate the gene.

As to the possibility of a drug that turned on telomerase so as to rejuvenate the body's tissues wholesale, Dr. Harley of Geron said, "It could be that is the ultimate upside, but we must move slowly on that."

Whether any of these hopes can be brought to fulfillment is an issue now overshadowed by several major doubts. Experts in aging and the biology of telomeres, while praising the new result, stress four outstanding limitations.

First, the body has many types of cells, including nerve and muscle cells, that never divide after the infant's development is complete and therefore never come up against the Hayflick limit—the finding that cells in test tubes divide about 50 times and then die, named after its discoverer, Leonard Hayflick. No one knows what causes these cells to age, but it is clear that telomerase is not going to reverse the process or have any effect on major diseases of senescence like Alzheimer's.

Second, if the telomere shortening system evolved in humans as a last-ditch defense against cancer, as seems probable, any intervention that paralyzes the system will postpone cell senescence at the high price of encouraging cancer.

Third, the aging of cells in the test tube is not necessarily the same as the aging of tissues in the body. Do people in fact die because their cells run out of telomeres?

Fourth, it is not clear that telomere shortening is the fundamental cause of cellular aging. Some other process of aging may emerge beyond the Hayflick limit.

The discovery of the human telomerase gene has at last allowed Dr. Harley to put the theory to rigorous test. Inserting this gene into human cells, along with a signal that forced the gene to become active, he and his colleagues found that the cells quickly built their telomeres back up to youthful length, and continued to divide long past the usual limit.

"Our results indicate that telomere loss in the absence of telomerase is the intrinsic timing mechanism that controls the number of cell divisions prior to senescence," the biologists wrote in their report.

Dr. Greider, who noted that mice regulate their telomerase gene very differently from humans, said the biologists' work "makes a strong statement that the telomeres have something to do with the life of the cell."

Dr. Leonard Guarente, an expert on cellular aging at MIT, described the research as "a beautiful piece of work" but said its relation to human aging remained to be defined.

Dr. Weinberg of the Whitehead Institute, who has recently entered the telomere field, said the finding was relevant to the senescence of human cells, but not necessarily to that of the human body. "If it were true that our life span is dictated by telomere shortening, then you would imagine that some humans die because their critical cells run out of telomeres, but there is no evidence for that at all. The Geron people are pushing this hard, but to my mind there is not much evidence for it."

Dr. Weinberg suggested that everyone might have enough telomeres to live to be 200 years old. What limits human life span is disease, especially cancer, and the telomere shortening system is a way of ensuring that the growth of incipient cancer cells is quickly halted.

"If you put telomerase into all our cells maybe they could live longer," Dr. Weinberg said, "but on balance it would be a very bad thing because it would no longer prevent those cells from growing without limits, and in the end malignancy is the greatest threat to human longevity."

—NICHOLAS WADE, January 1998

8

THE LATEST FROM THE FIELD

The twentieth century began with the rediscovery of Mendel's laws, welcomed at about its midpoint Watson and Crick's elucidation of the structure of DNA, and culminated in the decoding of the human genome sequence. But knowing the string of DNA letters in the human genome and understanding what they mean are two different things. It may take much of the 21st century to figure out all the steps that occur between the information embodied in the human genome and the creation and operation of the human body.

The human genome project began in 1990 with the goal of completing the last nucleotide by 2005. As the first article in this section relates, a finishing post of sorts was reached much earlier after a private company named Celera started its own human genome project in May 1998. The two rivals, Celera and the public consortium of academic centers, decided, after months of increasingly bitter exchanges, that their best interest lay in declaring a draw. At an elegant White House ceremony in June 2000, the two teams declared a joint victory, even though neither had completed the full genome, just a first rough draft.

But having nearly 3 billion units of DNA in hand was only a start. The two teams then turned to making first, preliminary interpretations of their data. As part of the peace pact that led to the White House announcement, the rivals agreed to publish their results at the same time, although in different scientific journals.

Perhaps the main surprise in their findings, as the second article reports, is how few human genes there seemed to be—a mere 30,000 or so, compared with the 100,000 that had been the standard estimate hitherto.

The human genome has been justified, both to government and investors, as a quest for new drugs and therapies. But of course the genome is far more than some particularly comprehensive medical textbook. Everything about inherited human nature, its darkest urges as well as its most altruistic, is encoded in the genome's four-letter alphabet. As the third article notes, biologists may discover that much more of our behavior than generally expected has a genetic basis.

Genetics is likely to unravel many other mysteries in the years ahead, including the many historical problems that can be addressed by working out who is related to whom. The long arm of DNA has reached back into history to corroborate the idea, long dismissed by historians, that Thomas Jefferson had children with his slave mistress Sally Hemings. DNA evidence has also been used to show the remarkable genetic continuity, at least as measured on the male side, of Jewish populations since the diaspora. Humankind was once a small population, maybe no more than 10,000 souls in all, but has since split into the thousands of different branches represented by today's ethnicities. DNA analyses can now trace these branches back to their trunk and examine how one is related to the other.

Genetics has never been more interesting, nor the pace of discovery so fast and furious.

—NICHOLAS WADE

Genetic Code of Human Life
Is Cracked by Scientists

IN AN ACHIEVEMENT THAT represents a pinnacle of human self-knowledge, two rival groups of scientists said today that they had deciphered the hereditary script, the set of instructions that defines the human organism.

"Today we are learning the language in which God created life," President Clinton said at a White House ceremony attended by members of the two teams, Dr. James D. Watson, co-discoverer of the structure of DNA, and, via satellite, Prime Minister Tony Blair of Britain.

The teams' leaders, Dr. J. Craig Venter, president of Celera Genomics, and Dr. Francis S. Collins, director of the National Human Genome Research Institute, praised each other's contributions and signaled a spirit of cooperation from now on, even though the two efforts will remain firmly independent.

The human genome, the ancient script that has now been deciphered, consists of two sets of 23 giant DNA molecules, or chromosomes, with each set—one inherited from each parent—containing more than three billion chemical units.

The successful deciphering of this vast genetic archive attests to the extraordinary pace of biology's advance since 1953, when the structure of DNA was first discovered and presages an era of even brisker progress.

Understanding the human genome is expected to revolutionize the practice of medicine. Biologists expect in time to develop an array of diagnostics and treatments based on it and tailored to individual patients, some of which will exploit the body's own mechanisms of self-repair.

The knowledge in the genome could also be used in harmful ways, particularly in revealing patients' disposition to disease if their privacy is not safeguarded, and in causing discrimination.

The joint announcement is something of a shotgun marriage because neither side's version of the human genome is complete, nor do they agree on the genome's size. Neither has sequenced—meaning to determine the order of the chemical subunits—the DNA of certain short structural regions of the genome, which cannot yet be analyzed.

With the rest of the genome, which contains the human genes and much else, both sides' versions have many small gaps, although these are thought to contain few or no genes. Today's versions are effectively complete representations of the genome but leave much more work to be done.

The two groups even differ on the size of the gene-coding part of the genome. Celera says it is 3.12 billion letters of DNA; the public consortium that it is 3.15 billion units, a letter difference of 30 million. Neither side can yet describe the genome's full size or determine the number of human genes.

The public consortium has also fallen somewhat behind in its goal of attaining a working draft in which 90 percent of the gene-containing part of the genome was sequenced. Its version today has reached only 85 percent, suggesting it was marching to Celera's timetable.

Today's announcement heralded an unexpected truce between the two groups of scientists who have been racing to finish the genome. Veering away from the prospect of asserting rival claims of victory, the two chose to report simultaneously their attainment of different milestones in their quest.

Celera, a unit of the PE Corporation, has obtained its 3.12 billion letters of the genome in the form of long continuous sequences, mostly about 2 million letters each, but with many small gaps.

A less complete version has been reported by the Human Genome Project, a consortium of academic centers supported largely by the National Institutes of Health and the Wellcome Trust, a medical philanthropy in London. Dr. Collins, the consortium's leader, said its scientists had sequenced 85 percent of the genome in a "working draft," meaning its accuracy will be upgraded later.

Both versions of the human genome meet the important goal of allowing scientists to search them for desired genes, the genetic instructions encoded in the DNA. The consortium's genome data is freely available now. Celera has said it will make a version of its genome sequence freely available at a later date.

In their remarks at the White House, Dr. Collins and Dr. Venter both sought to capture the wider meaning of their work in identifying the eye-glazing stream of A's, G's, C's and T's, the letters in the genome's four-letter code.

"We have caught the first glimpses of our instruction book, previously known only to God," Dr. Collins said. Dr. Venter spoke of his conviction from seeing people die in Vietnam, where he served as a medic, that the human spirit transcended the physiology that is controlled by the genome.

The two genome versions were obtained through prodigious efforts by each side, involving skilled management of teams of scientists working around the clock on a novel technological frontier.

Spurring their efforts was the glittering lure of the genome as a scientific prize, and a rivalry fueled by personal differences and conflicting agendas.

Dr. Venter, a genomics pioneer whose innovative methods have at times been scorned by experts in the consortium's camp, has often cast himself, not without reason, as an outsider battling a hostile establishment.

The consortium scientists were halfway through a successful 15-year program to complete the human genome by 2005, when Dr. Venter announced in May 1998 that as head of a new company, later called Celera, he would beat them to their goal by 5 years.

His bombshell entry turned an academic pursuit into a fierce race. Dr. Collins responded by moving his completion date forward to 2003 and setting this month as the target for a 90 percent draft.

"These folks have pulled out all the stops," he said of his staff in an interview last week. "They have achieved a ramp-up that is beyond anything one would have imagined possible."

The 15-year cost of the Human Genome Project, which began in 1990, has been estimated at $3 billion, but includes many incidental expenses. The consortium has spent only $300 million on sequencing the human genome since January 1999, when its all-out production phase began. Celera has not released its costs, but Dr. Venter said a year ago that he expected Celera's human genome to cost $200 million to $250 million.

The race opened with mutual predictions of defeat. The consortium's senior scientists predicted in December 1998 that Dr. Venter's method of reassembling the sequenced fragments of genomic DNA was bound to fail. In May 1999, Dr. Venter, confident of Celera's impending success, observed that the National Institutes of Health and the Wellcome Trust were "putting good money after bad."

The groups were divided by political as well as technical agendas. The consortium's two principal scientists, Dr. John E. Sulston of the Sanger Center in England and Dr. Robert Waterston of Washington University in St. Louis, insisted that the genome data should be published nightly, an unusually generous policy because scientists generally harvest new data for their own discoveries before sharing it.

Both of the consortium's administrative leaders, Dr. James D. Watson, and his successor, Dr. Collins, made a point of seeking out international partners so that the rest of the world would not feel excluded from the genome triumph. Thus even though centers in the United States and Britain have done most of the heavy lifting, important contributions to the consortium's genome draft have been made by centers in Germany, France, Japan and China.

Academic scientists have felt some chagrin that an altruistic, open and technically successful venture like the Human Genome Project should be upstaged by a commercial rival financed by the company that made the consortium's DNA sequencing machines.

But though Celera seeks to profit by operating a genomic database, Dr. Venter also believed that he could make the genome and its benefits available a lot sooner. He has succeeded in doing so, and in spurring the consortium to move faster.

Today's truce between the two teams offers several advantages. For Celera to claim victory over the consortium would risk alienating customers in the academic community. For the consortium, the surety of opting into a draw now may have seemed better than the risks of claiming victory with a complete genome much later.

Celera's version of the genome depends on the consortium's data. And the many small gaps in Celera's sequence will probably be filled by the consortium's scientists, adding further to their claim on credit for the final product.

The present truce between the sides is limited to today's announcement and an agreement to publish their reports in the same journal, although the details remain to be worked out. A joint workshop will be held to discuss the genome versions.

The versions of the human genome produced by the two teams are in different states of completion because of the different methods each used to determine the order of DNA units in the genome.

The consortium chose first to break the genome down into large chunks, called BAC's, which are about 150,000 DNA letters long, and to sequence each BAC separately. This BAC by BAC strategy also required "mapping" the genome, or defining short sequences of milestone DNA that would help show where each BAC belonged on its parent chromosome, the giant DNA molecules of which the genome is composed.

BAC's are assembled from thousands of snippets of DNA, each about 500 DNA letters in length. This is the longest run of DNA letters that the DNA sequencing machines can analyze. A computer pieces together the snippets by looking for matches in the DNA sequence where one snippet overlaps another.

But the BAC's do not assemble cleanly from their component snippets. One reason is that human DNA is full of repetitive sequences—the same run of letters repeated over and over again—and these repetitions baffle the computer algorithms set to assemble the pieces.

The stage the consortium has now reached is that all its BAC's are mapped, making the whole genome available in a nested set of smaller jigsaw puzzles. But the BAC's are in varying stages of completion. The BAC's covering the two smallest human chromosomes, numbers 21 and 22, are essentially complete. But many other BAC's are in less immaculate states of assembly. Many consist of assembled pieces no more than 10,000 units long, and the order of these pieces within each BAC is not known.

The sum of the assembled pieces in each BAC now covers 85 percent of the genome. This working draft, as the consortium calls it, is maybe not a thing of beauty but is of great value to researchers looking for genes and represents a major accomplishment.

Celera's genome has been assembled by a different method, called a whole genome shotgun strategy. Following a scheme proposed by Dr. Eugene Myers and Dr. J. L. Weber, Celera skips the time-consuming mapping stage and breaks the whole genome down into a set of fragments that are 2,000, 10,000 and 50,000 letters long. These fragments are analyzed separately and then assembled in a single mammoth computer run, with a handful of clever tricks to step across the repetitive sequence regions in the DNA.

The approach ideally required sequencing 30 billions units of DNA—10 times that in a single genome. Dr. Venter seems to have taken a considerable risk by starting his assembly at the end of March this year when he possessed only a threefold coverage of the genome. He has since raised his total to 4.6-fold coverage.

The decision may have been influenced by Celera's rate of capital expenditure—the company's electric bill alone is $100,000 a month—and by the need to sequence the mouse genome as well so as to offer database clients a two-genome package. The mouse genome is expected to be invaluable for interpreting the human genome, and Dr. Venter said today that Celera would finish sequencing it by the end of the year.

Because of having relatively little of its own data, Celera made use of the consortium's publicly available sequence data and, indirectly, of the positional information contained in the consortium's mapped set of BAC's. The consortium can justifiably share in the credit for Celera's version of the genome, another cogent factor in the logic of today's truce.

—Nicholas Wade, June 2000

Genome's Riddle:
Few Genes, Much Complexity

THE HUMAN GENOME IS the most precious body of information imaginable. Yet the biologists who yesterday reported their first analysis of the decoded sequence have found as much perplexity as enlightenment.

The chief puzzle is the apparently meager number of human genes. Textbooks have long estimated 100,000, a number that seemed perfectly appropriate even after the first two animal genomes were deciphered. The laboratory roundworm, sequenced in December 1998, has 19,098 genes and the fruit fly, decoded last March, owns 13,601 genes. But the human gene complement has now turned out to be far closer to genetic patrimony of these two tiny invertebrates than almost anyone had expected.

Dr. J. Craig Venter and colleagues at Celera Genomics report in this week's *Science* that they have identified 26,588 human genes for sure, with another 12,731 candidate genes. When they first screened the gene families likely to have new members of interest to pharmaceutical companies, "there was almost panic because the genes weren't there," Dr. Venter said.

Celera's rival, the publicly funded consortium of academic centers, has come to a similar conclusion. Its report in this week's Nature pegs the probable number of human genes at 30,000 to 40,000. Because the current gene-finding methods tend to over-predict, each side prefers the lower end of its range, and 30,000 seems to be the new favorite estimate.

The two teams, who discussed their findings in a news conference yesterday in Washington, found other oddities, too. Most of the repetitive DNA sequences in the 75 percent of the genome that is essentially junk ceased to accumulate millions of years ago, but a few of sequences are still active and may do some good. The chromosomes themselves have a rich archaeology. Large blocks of genes seem to have been extensively copied from one human chromosome to another, beckoning genetic archaeologists to figure out the order in which the copying occurred and thus to reconstruct the history of the animal genome.

As the modest number of human genes became apparent, biologists in both teams were forced to think how to account for the greater complexity of people, given that they seem to possess only 50 percent more genes than the roundworm. It is not foolish pride to suppose there is something more to Homo sapiens than Caenorhabditis elegans. The roundworm is a little tube of a creature with a body of 959 cells, of which 302 are neurons in what passes for its brain. Humans have 100 trillion cells in their body, including 100 billion brain cells.

Several explanations are emerging for how to generate extra complexity other than by adding more genes. One is the general idea of combinatorial complexity—with just a few extra proteins one could make a much larger number of different combinations between them. In a commentary in Science, Dr. Jean-Michel Claverie, of the French National Research Center in Marseille, notes that with a simple combinatorial scheme, a 30,000-gene organism like the human can in principle be made almost infinitely more complicated.

But Dr. Claverie suspects humans are not that much more elaborate than some of their creations. "In fact," he writes, "with 30,000 genes, each directly interacting with four or five others on average, the human genome is not significantly more complex than a modern jet airplane, which contains more than 200,000 unique parts, each of them interacting with three or four others on average."

The two teams' first scanning of the genome suggests two specific ways in which humans have become more complex than worms. One comes from analysis of what are called protein domains. Proteins, the working parts of the cell, are often multipurpose tools, with each role being performed by a different section or domain of the protein.

Many protein domains are very ancient. Comparing the domains of proteins made by the roundworm, the fruit fly and people, the consortium reports that only 7 percent of the protein domains found in people were absent from worm and fly, suggesting that "few new protein domains have been invented in the vertebrate lineage."

But these domains have been mixed and matched in the vertebrate line to create more complex proteins. "The main invention seems to have been cobbling things together to make a multitasked protein," said Dr. Francis S. Collins, director of the genome institute at the National Institutes of Health and leader of the consortium. "Maybe evolution designed most of the basic folds that proteins could use a long time ago, and the major advances in the last 400 million years have been to figure out how to shuffle those in interesting ways. That gives another reason not to panic," he said, in reference to fears about the impoverished genetic design of humans.

Evolution has devised another ingenious way of increasing complexity, which is to divide a gene into several different segments and use them in different combinations to make different proteins. The protein-coding segments of a gene are known as exons and the DNA in between as introns. The initial transcript of a gene is processed by a delicate piece of cellular machinery known as a spliceosome, which strips out all the introns and joins the exons together. Sometimes, perhaps because of signals from the introns that have yet to be identified, certain exons are skipped, and a different protein is made. The ability to make different proteins from the same gene is known as alternative splicing.

The consortium's biologists say that alternative splicing is more common in human cells than in the fly or worm and that the full set of human proteins could be five times as large as the worm's. Another possible source of extra complexity is that human proteins have sugars and other chemical groups attached to them after synthesis.

There's a different explanation of human complexity, which is simply that the new low-ball figure of human genes derived by Celera and consortium is a gross undercount. Dr. William Haseltine, president of Human Genome Sciences, has long maintained that there are 120,000 or so human genes. Dr. Randy Scott, chief scientific officer of Incyte Genomics, predicted in September 1999 that there were 142,634 human genes. Last week Dr. Scott said he accepted the rationale for the lesser number and now puts the human complement at around 40,000.

Dr. Haseltine, however, remains unshaken in his estimate of 100,000 to 120,000 genes. He said last week that his company had captured and sequenced 90,000 full-length genes, from which all alternative splice forms and other usual sources of confusion have been removed. He has made and tested the proteins from 10,000 of these genes. The consortium and Celera have both arrived at the same low number because both are using the same faulty methods, in his view.

"I believe their gene finding methods are far more imperfect than they own up to," Dr. Haseltine said, noting that 5 of the 10 genes in the AIDS virus were missed at first. "It's my personal conviction that as further studies of chromosomes continue the number of genes will rise until they match the number we project of 100,000 to 120,000."

Dr. Haseltine notes that the gene-finding methods used by the two teams depend in part on looking for genes like those already known, a procedure that may well miss radically different types of genes. His own method, capturing the genes produced by variety of human cell types, is one that Dr. Venter says in his paper is the ultimate method of counting human genes.

Dr. Haseltine is at present in a camp of one. Dr. Venter strongly disagrees, as do members of the consortium. Dr. Eric S. Lander of the Whitehead Institute last week challenged Dr. Haseltine to make public all the genes he had found in a 1 percent region of the genome and let others assess his claim. Dr. Collins said that there was "a terrific way to size up his claims—let an objective third party look at the data."

"I'd be glad to help arrange that," he said.

Dr. Haseltine said yesterday that he was contemplating the best way to respond and that he was "planning to do so in one form or another, in the open literature."

Turning from genes to chromosomes, one of the most interesting discoveries in this week's papers concerns segmental duplications, or the copying of whole blocks of genes from one chromosome to the other. These block transfers are so extensive that they seem to have been a major evolutionary factor in the genome's present size and architecture. They may arise because of a protective mechanism in which the cell reinserts broken-off fragments of DNA back into the chromosomes.

In Celera's genome article, Dr. Venter presents a table showing how often blocks of similar genes in the same order can be found throughout the genome. Chromosome 19 seems the biggest borrower, or maybe lender, with blocks of genes shared with 16 other chromosomes.

Much the same set of large-scale block transfers seems to have occurred in the mouse genome, Dr. Venter writes, suggesting that the duplications "appear to predate the two species' divergence" about 100 million years ago. He hopes that by sequencing the genomes of many other species he can reconstruct the history of the genome's formation.

Segmental duplication is an important source of innovation because the copied block of genes is free to develop new functions. An idea enshrined in many textbooks is that the whole genome of early animals has twice been duplicated to form the vertebrate lineage. There are several cases in which one gene is found in the roundworm or fly and four very similar genes in vertebrates. (The quadruplicated genes that failed to find a useful role would have been shed from the genome.)

But neither Celera nor the consortium has found any evidence for the alleged quadruplication. If this venerable theory is incorrect, the four-gene families may all arise from segmental duplication.

No one could expect a text as vast and enigmatic as the human genome to yield all its secrets at first glance, and indeed it has not done so. Dr. Venter said that the principal purpose of his paper was to describe the sequence and that he would convene conferences of experts to help further interpret it.

Dr. Lander said the consortium's analysis too was just preliminary. "We tried to write a paper that was not the last word on the genome but sketched all the directions you could go in," he said. "The goal was to launch a thousand ships, not to catalog a thousand genes."

—Nicholas Wade, February 2001

The Story of Us:
Other Secrets of the Genome

BIOLOGISTS HAVE TAKEN THEIR first look at the human genome and report that its 30,000 genes, though fewer than expected, will help decipher the genetic basis of many diseases and in time revolutionize medicine. But what will the genome tell us about human nature?

There the predictions become far less explicit, as if the genome were going to tell us everything about our bodies and nothing about our behavior.

Dr. J. Craig Venter, president of Celera Genomics, concludes his article about the human genome with a warning against what he sees as the dangers of determinism, "the idea that all characteristics of the person are 'hard-wired' by the genome." Dr. Francis Collins, leader of the public consortium that sequenced the genome and Celera's rival, said last week that "one of the greatest risks of this focus on the genome" is that people will draw the conclusion that their choices in life are "hard-wired into our DNA and free will goes out the window and we move into this mindset of genetic determinism."

It's easy to refute the advocates of genetic determinism, if any non-straw ones exist, because it is obvious to scientists that human behavior is not completely specified by the genome. But the opposite position—that evolution does not care about and has in no way shaped human behavior—seems equally implausible. Evolution is likely to have molded any and all behaviors that conferred survival value.

That may be easy to accept for behaviors like fighting and mating. But what about sophisticated human behaviors like art and religion? Art might seem a pure expression of human volition. But listen to paleo anthropologists debating the first appearance of art in the archaeological record and it seems that art, or some aspect of mind closely related to it, played a critical role in human evolution.

259

Most human societies have religious beliefs of one kind or another. Some may take that as proof of a universal creator. But for the biologist, the only question is whether such behavior is adaptive, meaning selected and genetically encoded because it conferred some survival value on the societies that possessed it. Religion is a source of spiritual values but has also served throughout history as a strong cohesive force among rival societies at each other's throats. Don't the bishops always bless the cannon?

"The human mind," writes the biologist Edward O. Wilson in his recent book *Consilience*, "evolved to believe in the gods. It did not evolve to believe in biology. Acceptance of the supernatural conveyed a great advantage throughout prehistory, when the brain was evolving. Thus it is in sharp contrast to biology, which was developed as product of the modern age and is not underwritten by genetic algorithms."

If a wide spectrum of human behavior from art to ethics is indeed adaptive, as biologists like Dr. Wilson believe may be the case, then the genome encodes a wealth of information about the essence of human nature. And that would transform it from a benign medical textbook into a document deeply disquieting to those who would prefer to believe the higher forms of human behavior and experience are transcendental states beyond the reach of mere genes.

Is it just human behavior that's exempt from evolution's shaping hand? Imagine the conversation among a group of fruitflies, whose genome was decoded last March:

"They'll never understand the Drosophilan nature from our genome— there's so much more to us than mere genes."

"Yes, the delirious high of being a wild young maggot, perpetually drunk while chewing through fermented fruit."

"And how could mere genes determine the thrill of the courtship as I sing to my inamorata by vibrating my wings, and she listens to my song and accepts me."

"Or rejects you with a buzz of her wings."

But dear Drosophilans, the alcohol dehydrogenase enzyme that lets you survive in alcohol is one of the best studied proteins in biology. And your courtship behavior, sophisticated as it may be, is tightly controlled by two genes that biologists call "fruitless" and "dissatisfaction." With a small change in the DNA of the fruitless gene, the males will produce an abnormal scent, court both males and females and fail to mate. With a different mutation, the males court only males and fail to sing their courtship song. Females with mutations in dissatisfaction resist mating by kicking wooers and flicking their wings to dislodge the suitors trying to mount them. How single genes can produce such surprising transformations is not yet fully understood, but it seems that the state of the fruitless and dissatisfaction genes pretty much specifies all the rules in the playbook of fruitfly dating.

With the human genome in hand, the way is open, at least in principle, to discover how it shapes the architecture of the human mind. "Nearly all human behaviors that have been studied show moderate to high heritability," write Dr. Robert Plomin and colleagues at the London Institute of Psychiatry in a commentary in the current issue of *Science*. They believe that, in time, "the human genome sequence will revolutionize psychology and psychiatry."

Dr. Plomin, who discovered the first gene that affects human intelligence, describes the influence of genes on human behavior as "probabilistic rather than deterministic." Dr. Wilson also expects that evolution has found it more effective to set loose prescriptions—epigenetic rules, he calls them—to guide those trying to survive in complex and hazardous jungle of human society.

All branches of human knowledge, he argues, from ethics to economics and aesthetics, will eventually be unified by understanding the genetic rules of the human mind. "The search for human nature," he writes, "can be viewed as the archaeology of the epigenetic rules."

So never fear—the human genome is nothing like the bland medical textbook that those who decoded it are intent on describing. When fully translated, it will prove the ultimate thriller—the indisputable guide to the graces and horrors of human nature, the creations and cruelties of the human mind, the unbearable light and darkness of being.

—Nicholas Wade, February 2001

Y Chromosome Bears Witness to Story of the Jewish Diaspora

WITH A NEW TECHNIQUE based on the male or Y chromosome, biologists have traced the diaspora of Jewish populations from the dispersals that began in 586 B.C. to the modern communities of Europe and the Middle East.

The analysis provides genetic witness that these communities have, to a remarkable extent, retained their biological identity separate from their host populations, evidence of relatively little intermarriage or conversion into Judaism over the centuries.

Another finding, paradoxical but unsurprising, is that by the yardstick of the Y chromosome, the world's Jewish communities closely resemble not only each other but also Palestinians, Syrians and Lebanese, suggesting that all are descended from a common ancestral population that inhabited the Middle East some four thousand years ago.

Dr. Lawrence H. Schiffman, chairman of the department of Hebrew and Judaic Studies at New York University, said the study fit with historical evidence that Jews originated in the Near East and with biblical evidence suggesting that there were a variety of families and types in the original population.

He said the finding would cause "a lot of discussion of the relationship of scientific evidence to the manner in which we evaluate long-held academic and personal religious positions," like the question of who is a Jew.

The study, reported in today's *Proceedings of the National Academy of Sciences*, was conducted by Dr. Michael F. Hammer of the University of Arizona with colleagues in the United States, Italy, Israel, England and South Africa. The results accord with Jewish history and tradition and refute theories like those holding that Jewish communities consist mostly of converts from other faiths, or that they are descended from the Khazars, a medieval Turkish tribe that adopted Judaism.

The analysis by Dr. Hammer and colleagues is based on the Y chromosome, which is passed unchanged from father to son. Early in human evolution, all but one of the Y chromosomes were lost as their owners had no children or only daughters, so that all Y chromosomes today are descended from that of a single genetic Adam who is estimated to have lived about 140,000 years ago.

In principle, all men should therefore carry the identical sequence of DNA letters on their Y chromosomes, but in fact occasional misspellings have occurred, and because each misspelling is then repeated in subsequent generations, the branching lineages of errors form a family tree rooted in the original Adam.

These variant spellings are in DNA that is not involved in the genes and therefore has no effect on the body. But the type and abundance of the lineages in each population serve as genetic signature by which to compare different populations.

Based on these variations, Dr. Hammer identified 19 variations in the Y chromosome family tree. The ancestral Middle East population from which both Arabs and Jews are descended was a mixture of men from eight of these lineages.

Among major contributors to the ancestral Arab-Jewish population were men who carried what Dr. Hammer calls the "Med" lineage. This Y chromosome is found all round the Mediterranean and in Europe and may have been spread by the Neolithic inventors of agriculture or perhaps by the voyages of sea-going people like the Phoenicians.

Another lineage common in the ancestral Arab-Jewish gene pool is found among today's Ethiopians and may have reached the Middle East by men who traveled down the Nile. But present-day Ethiopian Jews lack some of the other lineages found in Jewish communities, and overall are more like non-Jewish Ethiopians than other Jewish populations, at least in terms of their Y chromosome lineage pattern.

The ancestral pattern of lineages is recognizable in today's Arab and Jewish populations, but is distinct from that of European populations and both groups differ widely from sub-Saharan Africans.

Each Arab and Jewish community has its own flavor of the ancestral pattern, reflecting their different genetic histories. Roman Jews have a pattern quite similar to that of Ashkenazis, the Jewish community of Eastern

Europe. Dr. Hammer said the finding accorded with the hypothesis that Roman Jews were the ancestors of the Ashkenazis.

Despite the Ashkenazi Jews' long residence in Europe, their Y signature has remained distinct from that of non-Jewish Europeans.

On the assumption that there have been 80 generations since the founding of the Ashkenazi population, Dr. Hammer and colleagues calculate that the rate of genetic admixture with Europeans has been less than half a percent per generation.

Jewish law tracing back almost 2,000 years states that Jewish affiliation is determined by maternal ancestry, so the Y chromosome study addresses the question of how much non-Jewish men may have contributed to Jewish genetic diversity. Dr. Hammer was surprised to find how little that contribution was.

"It could be that wherever Jews were, they were very much isolated," he said. The close genetic affinity between Jews and Arabs, at least by the Y chromosome yardstick, is reflected in the Genesis account of how Abraham fathered Ishmael by his wife's maid Hagar and, when Sarah was then able to conceive, Isaac. Although Muslims have a different version of the story, they regard Abraham and Ishmael, or Ismail, as patriarchs just as Jews do Abraham and Isaac.

—Nicholas Wade, May 2000

Appendix: A Guide to the Language of Biology

THE ARTICLES in this book were written to be self-explanatory, but the reader may nevertheless find helpful a brief description of genes and how they work. In this outline, the technical terms commonly used in this book are marked in boldface type.

THE CELL AND ITS EQUIPMENT

The basic unit of living things is the **cell.** Many organisms, such as bacteria, consist of single, free-living, independent cells. Animals are constructed of many cells that have learned to live together. The human body consists of about 10,000 trillion cells. Most are about one fifth the size of the smallest object the eye can see.

Every cell has an outer coat, or membrane, with special pores for imports and exports. Fuel such as glucose may be imported, signaling agents such as hormones may be exported. Inside the cell an array of chemical activity goes on. The glucose is broken down and converted to energy, which drives all the other processes.

The inside of a cell is filled with a liquid, mostly water, that is called the **cytoplasm.** The cell also possesses various specialized structures. Animal cells have a **nucleus,** an interior compartment where the cell's own DNA is kept. In the cytoplasm are many **ribosomes,** miniature factories where protein molecules are assembled. There are also **mitochondria,** chemical plants that handle the cell's energy needs. Mitochondria are thought to have once been free-living bacteria that were captured by animal cells. They possess their own DNA, known as **mitochondrial DNA.**

The **protein molecules** manufactured by the ribosomes are of many sorts. They are the cell's working parts. Structural proteins give the cell rigidity and enable it to move when necessary. Transporter proteins usher chemicals in and out of the cell, and from one compartment to another. **Enzymes** manage all the cell's chemical reactions, bringing together the necessary chemicals to make a desired product. A special suite of enzymes manages the DNA, copying it when the cell divides and proofreading the new copy for errors.

Bacteria work on much the same principles as animal cells except that they do not have mitochondria, nor do they possess a nucleus.

GENES AND CHROMOSOMES

The information needed to operate the cell's chemical factories and everything else it does is programmed into its DNA. The DNA is organized into long molecules known as **chromosomes.** Ordinary human cells possess 23 pairs of chromosomes. One member of each pair comes from a person's father, one from their mother. The twenty-third pair is known as the **sex chromosomes** because they determine an individual's sex. Females have two **X chromosomes**; males have one X chromosome and one **Y chromosome.** X and Y are names that come from the chromosomes' shapes under the microscope. The sperm and egg cells possess a single set of chromosomes, which are pooled at fertilization.

Each chromosome consists of a single DNA molecule, typically 100 million units in length in the case of humans. The molecule is wound round special spools called **nucleosomes,** which keep the DNA tidy but also accessible when needed.

Chromosomes have a region known as a **centromere** in their middle and **telomeres** at their ends. The centromere is DNA of a special sequence which is recognized by the machinery that pulls each duplicated chromosome apart when the cell divides. The telomeres protect the tips of the chromosomes from fusing with each other. They also count the number of times a cell divides, and force it into senescence after it has divided its allotted number of times.

DNA is composed of chemcial units known as **nucleotides,** which are strung together in a long chain. There are four different kinds of nucleotides, called **adenine, guanine, thymine** and **cytosine** and known for short as A, G, T and C. The order of these chemical letters spells out the message that is the organism's genetic information.

The DNA molecule consists of two parallel chains of nucleotides that spiral around each other like the two sides of a spiral staircase. The steps of the staircase are formed by sidechains of opposing nucleotides. The nucleotides pair up in only two ways, A with T and G with C. When the cell divides, each DNA molecule is split into two chains, and a new partner is made for each chain. The structure of the old chain determines the structure of the new chain through the pairing rule that A will bind only with T, G with C and vice versa. The nucleotides are also known as **bases,** and a complementary pair of nucleotides as a **base pair.** Human cells have three billion base pairs of DNA.

The information in DNA is stored in larger units known as **genes.** Each gene contains the information to specify the manufacture of one protein. Humans are thought to have about 100,000 genes.

Genes specify proteins through a relationship called the **genetic code.** Proteins are made of units called **amino acids,** of which there are 20 different kinds. Although proteins come in all different shapes and sizes, they are generally made of a single chain that folds up in a specified way. A string of nucleotides in DNA corresponds to the string of amino acids in a protein. The correspondence is via a triplet code, meaning that a set of three nucleotides specifies one amino acid.

Since DNA is confined to the nucleus, and proteins are made by the ribosomes in the cytoplasm, how does the information pass from one to the other? The cell makes a copy of the gene, a process known as **transcription.** The transcribed copy is in the form of a molecule called **messenger RNA** because of its role in conveying the genetic information. As the messenger RNA ratchets through the ribosomes, the protein chain it specifies is generated in lockstep.

RNA is chemically similar to DNA except that its backbone is made of a more flexible unit, known as ribose, whereas DNA has deoxyribose for its backbone. The sequence of the RNA is complementary to the DNA of the gene, obeying the same pairing rules as the two DNA chains. However, in place of thymine, RNA has a base known as **uracil,** designated U for short. When a messenger RNA copy is made of DNA, U pairs with A, G with C.

In the genetic code, the sequence AUG in messenger RNA specifies the amino acid known as methionine, CAC specifies histidine and so forth.

When the cell copies a gene and makes the protein it specifies, the gene is said to be expressed.

MUTATIONS AND EVOLUTION

A **mutation** is the name given to any change in the ordinary form of a gene. The simplest kind of mutation is the change of one base for another. Sometimes bases are deleted; sometimes extra bases are inserted.

The change of a single base can cause a different amino acid to be specified at that point in the protein. For example, the mutation of GAG to GTG, a one-base change, alters the amino acid being ordered from glutamine to valine. Depending on the importance of the amino acid's role in the protein, the effect of the change may be anywhere between harmless and devastating. This particular change, in the beta chain of the hemoglobin molecule, is the one that causes sickle-cell anemia.

Sometimes a single base is added or deleted. These mutations are much more serious because from that point on in the messenger RNA chain, the ribosome reads the wrong set of three bases. Phase shift mutations, as they are called, produce the wrong amino acids throughout the rest of the protein chain, with the result that the protein is usually dysfunctional.

Most mutations are harmful, many are neutral, and a few are beneficial. This spectrum of possibilities offers **evolution** its chance to shape new genes and ultimately new species. Organisms carrying the beneficial mutations survive and reproduce, and the mutation becomes more common in subsequent generations.

THE THREE KINGDOMS OF LIFE

Very early in the evolution of life on earth, living cells became divided into three kingdoms known as **Prokarya** (bacteria), **Eukarya** and **Archaea.**

Bacteria are single-cell organisms that have no nucleus. Archaea are also single celled, and were distinguished from bacteria only recently. The Eukarya have a nucleus. Most of the Eukarya too are single-cell organisms like amebae. A relative handful have learned how to form multicelled organisms. This handful includes all animals, plants and fungi.

Since there is only one tree of evolution, all organisms and species are ultimately related to each other. Biologists like to trace this relationship in terms of genes. The gene that makes insulin in humans closely resembles the insulin gene in pigs. It resembles less closely the insulin gene in whales, and even more distantly that in the roundworm. A **family tree**, or **phylogeny,** of the insulin gene can be constructed, leading back to some ancestral insulin gene, and the tree will pretty much mirror the evolutionary relationships of the animals involved.

The phylogenetic relationship among genes is extremely helpful in recognizing the function of new genes that have been identified only by their DNA sequence. A new DNA sequence can be compared against all those recorded in DNA data banks; a computer program will list all similar genes that have been found, along with a measure of the extent of similarity, or **homology,** and the identity of the homologous genes, if known.

One group of entities that does not fit well into any of the three kingdoms is that of viruses. Viruses are not capable of independent existence outside of the cells they infect and are not alive in the same sense that cells are. A virus is a cluster of errant genes that has learned to parasitize certain cells. Viruses have afforded the keys to understanding many aspects of how cells work.

GENOMICS

Genes used to be studied one at a time, but with the invention of DNA sequencing machines it has become possible to consider the total DNA of an organism, usually referred to as its **genome.**

Study of the sequence and structure of the genome is called **genomics.** Understanding what the sequence means is known as **postgenomics.**

The genomes of many bacteria consist of a single, circular chromosome. Human and other animal cells have linear chromosomes. An important feature of animal genomes is that much of the DNA does not code for genes. The **noncoding DNA,** also known as **junk DNA,** consists mostly of the same few sequences repeated over and over again. The purpose of the noncoding DNA, if any, is not understood. As much as 97 percent of human DNA is noncoding.

A second unusual feature of animal genomes is that the genes do not occupy continuous stretches of DNA, but consist of coding regions separated by noncoding DNA. In a somewhat confusing terminology, the coding regions of a gene are called **exons** and the interspersed regions are called **introns.**

When the gene is first copied into RNA, the cell has special splicing machinery that removes the introns and joins the exons together as finished messenger RNA.

The reason for this curious system is thought to be that exons can more easily be mixed and matched over the course of evolution so as to form novel genes.

Bacterial genomes are far more compact than eukaryotic genomes. They have very little noncoding DNA and no introns.